S0-DVE-486

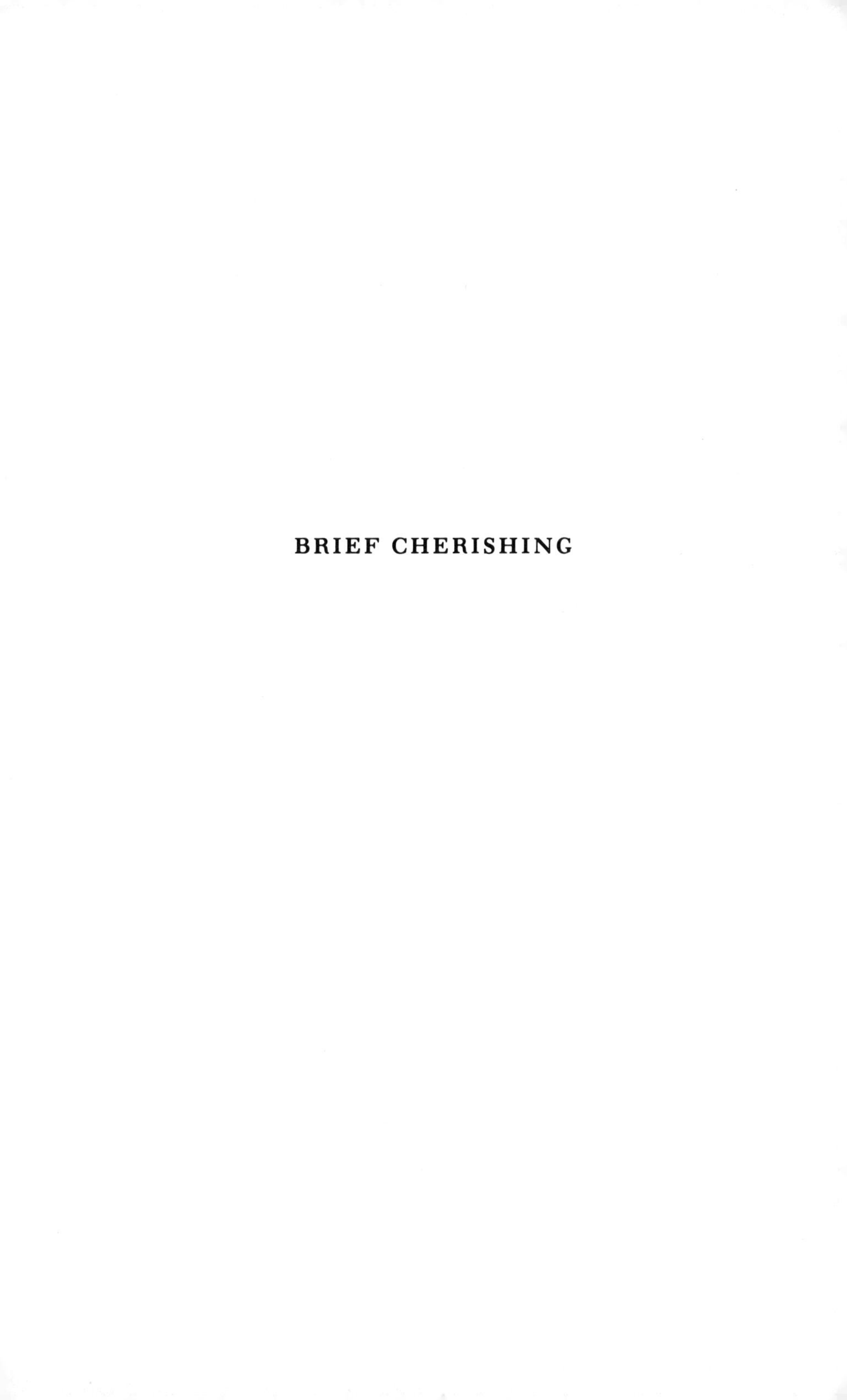

BRIEF CHERISHING

BRIEF CHERISHING

A Napa Valley Harvest

HILDEGARDE FLANNER

Wood Engravings by Frederick Monhoff
Foreword by Janet Lewis

John Daniel, Publisher
SANTA BARBARA
1985

"Miguel" was first published in *Confrontation*.
"Jacinto" was first published in *The New Yorker*.
"The Old Cherry Tree" was first published in *Westways*.

Design by Jim Cook
Typography by Cook/Sundstrom Associates
SANTA BARBARA, CALIFORNIA

LIBRARY OF CONGRESS CATALOGUING IN PUBLICATION DATA
Flanner, Hildegarde, 1899-
BRIEF CHERISHING
1. Flanner, Hildegarde, 1899- —Biography
2. Poets, American—20th century—Biography.
3. Napa River Valley (Calif.)—Social life and customs.
I. Title
PS3511.L28Z464 1985 811'.52 [B] 85-16107
ISBN 0-936784-03-2 (PBK.)

Published by
John Daniel, Publisher
POST OFFICE BOX 21922
SANTA BARBARA, CALIFORNIA 93121

CONTENTS

FOREWORD

I HAVE ALWAYS read Hildegarde Flanner with great delight. She writes with wit and freshness, "an unpretentious grace," an ease of colloquialism and an extraordinary sense of the language, of centuries of culture—oriental as well as western—behind her natural way of speaking. Tender, and never sentimental, very sharp, and wonderfully indignant at the foolish behavior of the human race, and greatly sad at the tragedy of some human choices—like war.

"Miguel," "Jacinto," and "The Old Cherry Tree" in this new and regrettably small collection charmed me as much as her memoirs of the Berkeley fire of 1924 and her garden in Altadena before the days of the great population invasion. *A Vanishing Land* also, published by John Daniel under the imprint of No Dead Lines, is fine reading and as potent a comment for us now on the too rapid changes in our world. However, to say that the long section called "A Brief Cherishing" delighted me is not suitable. It is delightful, yes, but much more. It is a vivid evocation of some of the best years of a long and deeply happy marriage, the story of a great love, and a great experience of living on this loved earth. It could be called unself-conscious in that it is not concerned with presenting a self, but is a clear and tender and witty vision of life perceived, fortunately remembered and recalled for us. A very beautiful and enlivening book.

—JANET LEWIS

MIGUEL

IT WAS MY BIRTHDAY. I would not have asked for the unexpected gift that suddenly appeared in the garden of our home in northern California.

When we sold our old home and garden in southern California and came north to live, we settled on a ranch property in a vineyard valley where my husband designed and built a new house. We had wanted to "spread out" from the restrictions of a suburban lot. Americans do not have the talent, so exquisitely practised by the Japanese, to make a suburban lot look like an unlimited landscape. So we spread out. While my husband established a water system, built a deer fence, and laid down the architectural design of premises and a garden I joined him in planting shrubs and trees. This called for help.

We were not used to working with Mexicans, but found them friendly toward plant life in a way that attracted and surprised us. They did not despise plants as workers in this country often do. Scarcely ever does a Mexican gardener behead, strangle, or otherwise execute a plant because it is an obstruction. In addition to that advantage of restraint in character they were often owners, especially the Guatemalans, of quite beautiful hats with low crowns, and brims rolled to a perfection that showed many generations of craft and style. These hats always suggested to me that they were woven, not to rest on the heads of hot labor but of cool aristocratic leisure. Yet it was only right, I told myself, that grace and fashion would reward the stooping back and the weary shoulders.

However, I found a drawback under the hat, which was a tendency in one Guatemalan boy working for me, to extract each weed individually from the earth, give it a shake, place it carefully on a slowly accumulating pile, and deliberate before the next neat deracination. When I pointed out that this was costly work he said, "It is the only way to do it." And it was the only way he did it.

There was another group, not natural gardeners, but intelligent and willing to work. Local citizens referred to them suspiciously as "those hippies." But they were not the same, although like the typical hippies they displayed beards and beads and the usual colorful regalia that hung and swung about them. I did not mind that at all. I enjoyed the approach through the garden of a bearded figure in a long red cape that swept the earth. A long red cape gives a stride of caravans and desert distances to the first moments of the day's work. The approach of a pair of blue jeans provides no such drama. Some of these men were young college drop-outs. One of them had dropped out of an instructorship. They found the academic life narrow, and chose instead the disciplines of a mountain-forest commune with echoes of Oriental religion. I took a ride in their old pick-up on an errand, and saw that the windshield was a shrine with OM MANE PADME HUM painted at the top, and a branch of my own bamboo hung alongside for its ancient meanings. I was startled to find myself looking into a photograph with intense mystical eyes of a dark man who piercingly stared back at me. The Guru.

This group had interests that took them from gardening eventually. In time various workers came to us, and prominent among them was the one who had suddenly and unexpectedly appeared around the corner of the courtyard on my birthday. I do not try to make any significance out of the circumstance that it was this particular day on which Miguel Torres y Rivera returned. Yet as I review the event it seems to me that it added unduly to my age, which is already too much. Miguel was one of the Mexicans best remembered who had come to work for us. I had not seen him for nearly two years when he had reluctantly left this valley and its welcome wages to return to Mexico and his native Michoacan. Now, on this particular morning, he inquired politely about my health and in answer to my murmured question assured me that he was also, *bueno, gracias*. It was the same ritual with which all the days of work had formerly begun. Now he added that I was to call for him the next morning, *domingo*, Sunday, at the *casa* of Senora Mendoza in the town nearby. In my surprise, and having lost in two years most of the very little Spanish anxiously and poorly acquired to deal with him originally, I was unable to tell him that I had a good *hombre* (who, *¡gracias a Dios!* spoke English), and that I did not need more help, and, with regrets, had no work for Miguel to do. All this I tried, in a wrestling bout between two languages, to convey. It was evident that he understood. Oddly, the information seemed to be beside the point. He added a few words I comprehended only later, repeated,"*las ochos*," eight o'clock, and was gone. I ran to look. A car was disappearing down the hill.

Miguel was one of the many Mexican laborers who manage, one way or another, to arrive somewhere in the United States and, legally or not, stay for a while to work until the Immigration Service catches up with them. In the meanwhile they have labored in orchards, vineyards, cotton, vegetable and rice fields and ranches or suburban gardens. My own garden is a country garden and Miguel had easily moved in upon its seclusion through a telephone call to me from the kindly Mexican-American woman who runs a hotel and apartment house for workers in our area. Mrs. Mendoza is careful about her recommendations, but her own sympathies are involved in trying to find employment for the most needy. "He is nice boy, very nice man," she says, "I know he is nice one." So it was that

Miguel had become, for a considerable time an attachment to our garden, our woods and our concerns.

There exists between workers and the government, as is well known, the onerous operation of employment taxes, and in the case of the Mexican worker the system starts when he has obtained a card that legalizes his temporary residence. I had enquired of Sra. Mendoza whether Miguel had a card. "Jesus Christ, honey," she cried, "they don't none of them have no cards." Well, perhaps this was taking a chance, but in many cases employment was brief, and the figuring at day's end was a fatigue of bookkeeping. Mine was sure to be bad. The government was better off without it.

Miguel's unexpected return and quick disappearance caused me to begin remembering many things about him. Quite helplessly I began to remember and could not stop. One incident after another swam into my mind. One hears of that uncanny rush of recollection that is said to occur to a drowning man in which his past life is concentrated and, in one dry flash unrolled before his flooded eyes. I would be adrift to suggest that my acquaintance with Miguel had more than expendable importance in my life. Still, there had been odd pleasures, and mild crises which would always float in my mind, I felt, and without much hindrance, as now, rise to the surface of my memory. Unfortunately for me I incline toward an exaggeration of remedial and custodial responsibility for people who work in my garden. In regard to Miguel the situation required a mite too much from my good-will volunteerism. It seemed all the time a matter of hostile bugs that had bitten him, a childhood habit of sore throat, a mysterious ear-ache, colds, coughs, indefinite and wandering fevers, strange lassitude, sprains, small wounds from agave points—which he should have avoided, being, like the agaves, a Mexican himself—and most of all, poison oak. Between Miguel and poison oak there was a raging affection. He went toward it as toward a sweetheart. So it was that medical ministrations of one kind or another were as common as snail bait. My husband, Fred, who rarely asked for explanations of anything at all merely replenished the bandages, iodine and unguentine. His occasional smile meant, "Don't overdo it." Our son, John, with sour good judgment remarked, "That Mexican from Michoacan is just looking for an extra mother." Miguel's vivid and wistful descriptions of his mother's garden and its tropical fruits sounded, indeed, like a safe maternal *paraiso,* or paradise. And suddenly, remembering this, the words he spoke hastily as he left after his unforseen arrival came back to me. Now they were clear, translating themselves and shocking me. "*No hay dinero por comida.*" *I have no money for food.* Yes, he was thinner. Perhaps he had been living on tomatoes. Working in the tomato fields was the only work he could find to do at home, he had told us. And yet, in spite of poverty and the aggressive host of small troubles that bit and beset him he was never morose. He was almost a merry person.

One of my chief memories had been of his singing. He was a great singer. By "great" I mean constant, not appealing nor melodic. Just non-stopping. Much breath, many words, and apparent emotion flowed from him, but I never detected a coherent tune. "Of what do you sing, Miguel?" He regarded me with reproof. "Love, *naturalmente,*" and he would warble on and on. I would get a meaning, much as one gets the scent of pine needles even though the branches are out of reach. I asked him if he had anything to read at night. "No," he said with dignity. "I spend my evenings writing poetry." Ah! This was very exceptional news. "And what do you write about, Miguel?" Again the same look of reproof. "Love, *naturalmente,*" I said apologetically.

Miguel was, as well, a good whistler and his entertainment in this accomplishment pleased our grandchildren. It was obvious that he regarded the grandchildren covetously. He had told me, at the beginning of our acquaintance, "*Cuatro años de matri-*

monio, y nada, nada." ("Four years of marriage and nothing, nothing.") He spoke sadly, then whistled like a canary in a thicket of nightingales until he stopped to laugh with the delighted children.

As Miguel had no automobile I had to pick him up whenever he came to work and take him back in the evening to Sra. Mendoza's apartment house. En route there might be a long stop at the local market while I waited for Miguel to do his shopping, and he would emerge with abundant supplies of tortillas, tomatoes, onions, chilis, sauces, and an extra stack or two of tortillas on top of everything else needed for a good Mexican dinner. I could roughly calculate how much of my money lay in the big sacks and how good, with such ingredients, money can smell. I also became tired of hauling him and the tortillas back and forth and suggested, rather churlishly I suppose, that he buy a bicycle. "Never!" he cried. "Why not?" "*La bicicleta* is very dangerous. I would fall off." "*Naturalmente,*" I said.

Miguel, like the other Mexicans, was good with plants. He was never indifferent, but appeared to hope personally that whatever he put into the earth would enjoy its roots there in comfort, peace and human affection. I always made a list of chores for him to do and carried it with me as he had an abrupt way of shouting, "I'm through! What next?" He would even shout, "What next?" before commencing his first job. I found this upsetting and was glad that there was occasionally work for him to do out of sight. In our woods is a shrub locally called broom, known by us as Genista. Botanically it is Cytissus and an almost indomitable nuisance. It is a legume but the deer, in their usual inability to benefit the land-owner—(even their droppings are too dainty to enrich the soil)—traipse through it indifferent to foliage and pods or fly above it in their lovely, effortless leaps and never bother to consume a half-inch leaf. It blooms happily and heavily with a small yellow, fragrant pea-blossom and sets seeds by the million that remain viable into eternity. It is a bad thing to have around, but we seem to be the only people who fight it. Like so many introductions or escaped plants it suppresses native vegetation. Wishing that Miguel were a flock of hungry goats, (goats, without stopping to think, will eat it for a while), we turned him loose in the woods where he stayed for days at a time, pulling up the unwanted shrub, roaming under the redwoods and tall fir trees and suddenly breaking into a greeting of nightingale music if the grandchildren came up to sit in the tree-house their father had built for them on top of a huge redwood stump. One afternoon I climbed the steep road to the woods to see what he was accomplishing. I was too late, as he was already descending. This, before I saw him I soon knew by the hymn to love that floated ahead of him as he came down, water-bottle in one hand, lunch-box and hoe in the other, skipping on the tips of his feet and never slipping on the loose pebbles of the sharp road.

Miguel regarded my car, a nice yellow Maverick, with much envy. He could not hope to have a car of his own and he longed to wash mine which was always covered with dust from the ranch road that leads to our house. I parked it beside a stand-pipe and warned him not to use too much well-water. He took a long time, what with music-making, polishing and admiring his work. When he was through the Maverick shone as though it had just rolled from Mr. Birleffi's show room. Proudly I got into my car and expected to back away in style from the stand-pipe. But the wheels spun under me and the bright, clean vehicle began to sink in soft, wet mud. To get it out of the mud and get the mud off the wheels and wherever it had landed without making more mud was an enterprise that caused Miguel no grief at all, but rather, several kinds of ill-assorted mirth and a new kind of whistle that indicated, I presumed, astonishment. He could not understand my disgust at the waste of water, time and wages.

Usually there were only routine things to do, take the runners off the strawberries, take the hastily growing purslane out of the petunias, rake the paths, and where did you leave the rake, *el rastro,* the last time? One day, which was to be a day of significance, Miguel had barely got started when, immediately and with a pronounced commotion, he found a snake. We have few snakes, or those we have are not often in sight. I had not seen one for many months. There had been no opportunity to introduce one of our snakes to Miguel, or the other way around. I was surprised at his loud reaction to the first snake. Perhaps it was due to a sensible fear of the arboreal snakes of the tropics. He came running toward me with something twining and writhing on the handle of his shovel and shouting, "*Culebra! ¡Muy venenosa!*" And he made a motion above the snake, which had dropped to the ground. "*Voy matar!*"("I'm going to kill it!") "No, no!" I cried, "It's not poisonous. It's my friend. *¡Amiga!*" I had seen at once that it was the beneficial King Snake. Miguel looked at me in disbelief, and when I stooped and picked up the snake in my hands, to show him that it was not dangerous, he backed away in horror. It was a very fine Mountain King Snake, banded in black, yellow and red. It lives on rodents and has the reputation, as do all King Snakes, of attacking rattlers. To convince Miguel that the snake was harmless to man—and to me—I put it across my shoulders and it hung docilely around my neck, its three feet or more of beauty rippling slightly. Then I put it down near a dry stone wall where I assumed it might be living, and it disappeared neatly between the stones. Miguel himself, looking ill, disappeared as well after giving me a frightened glance that plainly said he thought me out of my mind. Not many hours later, whether in bravado or with the wish to assert his nonchalance, he turned up with something else discovered in his work, this time three evil-looking scorpions. "*Mis amigos,*" he said with a grin, and chuckled at my evident distrust of them. "*No venenos.*" He tipped the awful little creatures from one palm to the other. They strutted in his hand with their tails up over their backs, as they had been parading for centuries in the Zodiac and through the Pharaonic arts of Egypt and the symbolism of many cultures. The scorpions in this valley are not deadly, but their sting can be infectious, and any scorpion that stung Miguel would sting him twice as hard as it would sting anyone else, and it would take more than one extra mother to deal with the consequences. "Do you want me to kill them?" I stared at the scorpions with dislike. Far, incalculably far behind them and through the mists of ignorance that lie in my mind I could discern the magical vibrating gold of the scorpion on the head of Selket, one of four goddesses guarding the tomb of Tutankhamun, and then the haunting perfect head of the young ruler himself, bound with a chaplet of gold that bore a stately scorpion of gold rearing in its ordained mystery above the Pharaoh's brow. In wonder and humility I gave an impossible command. "Do not kill them but, *por favor*, put them back where they came from."

Miguel's concept of his work had seemed to include not only respectful care for his employer's plants, and the pleasant vaudeville of singing and whistling, but something else I had nearly forgotten, but when I did remember it I did so in detail. It was the grooming of his person. It sometimes amounted to an imaginary ceremony in which I, as an impatient audience of one, partook unseen by the priest of the mystery himself. One day as I glanced out of the window I had seen him standing in the sunlight with his comb in his hand and slowly combing his hair. He appeared to be gazing, with pleasure, into a nonexistent mirror. He gave his head a toss. His hair was very short and there was absolutely nothing to toss, but this motion was repeated as if for a ritual arrangement of long, heavy hair. With a circular drooping motion of his head he swung an almost visible pelage from one side to the other, swept his head up again, proudly, and parted once more all that hair that had, *naturalmente,* got superbly mussed up. When this foolishness finally ended I considered how fortunate I was. I did not have to go all the way to San Francisco to see a skilled exhibition of convincing pantomime.

There were times when I heard my bamboo windbells ringing and ringing. The word is right for bells, but for this unique Oriental instrument not quite appropriate, for the sound that is made is a shallow orchestral clatter. It suggests wind blowng through a bamboo grove, and if you happen to be familiar with bamboo and to have collected some of its many kinds and lived with them for years, as I have, it is a sound that is cherished. But if no wind is stirring to make the delightful clatter, then *naturalmente,* it must be Miguel standing there barely causing the hollow culm-lengths to swing and knock against each other. This was what I hired him to do? But I was touched by the wistful satisfaction on his face. He was thinking of his family, the wife to whom he wrote the poems and to whom he would ask me to address his envelopes for letters at long intervals. Song by day, poetry by night left no need for rude communication in poorly written letters. "Wouldn't you be better off at home with your family where it is so much cheaper to live?" I had asked him. "It is not cheaper to live where there is no money to live on." This was the stark logic I had learned from his sober Spanish truth.

It had cost him two hundred dollars, I learned later, to return to our valley to get work. So many others had also come that work was very scarce. I knew that he would hope to work as frequently as I would take him. That meant I must manage—and pay—two gardeners, one of whom I did not need and one who was one more than I needed or could afford. With a sigh I picked up the stout Cassell's Spanish-English Dictionary I had bought for the vocabulary struggles of his earliest employment, and also the notes I had kept. Again I nervously faced the fact that a long column of very common Spanish verbs are crankily irregular. So musical a language, and so mean. Also I began to ask myself about the eerie twilit indecisions that afflict even the common daylight use of language in general. So many meanings were only demi-meanings, slippery statements of contradiction and self-denial where possibilities, threats and abraded truths hug each other weakly and desperately. *Perhaps, in spite of, not at all, whenever, although, however, in as much,* as I stop to consider these and their kind I long for simple, pure profanity to hang onto. And how had the earliest speakers among the peoples of the earth ever managed to spit the roughage and toughage of trial meanings from their sore tongues and finally draw up to the board of clear communication? For a long time they must have choked on *might, would, should, ought* and acid morsels of that kind without whose intentional chemistries of clarity a battle could be lost, a tribe extinguished, a choice woman thrown to the wrong man. Fascinating problems, but I had no learning to deal with them, and even too little time to explore the paunchy volume of Cassell for the few practical words I should use in my garden. Miguel had returned. Did I need him? *No.* Did I have to hire him? *Yes.* Why? *Don't be estúpida.* Tomorrow I must be ready to greet him as it had been before, to assure him with thanks that I am well, to hope that he is well also, and that his family is well. I must be ready, before I sleep tonight, to meet the frustrations of trying to be simple and explicit in the midst of very fancy confusion. I spent the evening in harassment, rather than practice, of the verbs to put, to bring, to carry, to say, to go, to be, to remove, to wish, to prefer and how many others whose challenging irregularities left me sad and cringing. My husband observed my predicament. "You can't learn a language overnight," he advised me kindly. "You are exhausting yourself. Forget that Mexican." "What shall I tell him to do first?" I implored. I went to bed with the verb *fertilizer* (easy to remember), twelve shrubs of *Camellia sasanqua* and a bucket of cotton seed meal. I woke early, rehearsing the preterit of *decir*, had a hasty breakfast, got some hoses going *(mangueras)*, put some tools in the wheelbarrow *(caratilla)* and took off for town in my dusty Maverick to Sra. Mendoza's house (*casa*, of course.) There I waited in my car for what seemed a long time. No Miguel. I had leisure to admire the senora's roses and bright red geraniums, and wonder what kind of grapes she had on her old pergola. I counted her hollyhocks. I contemplated the two large chairs visible in her patio. They had come from a beauty parlor, without doubt, for one of them had the remains of a hair dryer attached to the back. Then I lifted my voice. "Miguel!" three times. At last a second story window was opened. "You waiting for Miguel?" I did not know the voice. "Yes, I am waiting for Miguel." "He's gone." "Miguel's gone? Gone where?" "To the big valley on a bus to pick apricots." The window was closed.

Ahhh. . .I relaxed. With a quiet bang all the tools fell from my hands and the chores lay down in my mind. In this glorious second I was set free from the struggle to find the correct words in Spanish that would set the day to rolling in a proper sequence without embattled grammar, scandals of wrong vocabulary, and sudden strangling silence, while finding chores for the starving Miguel to perform between swinging his long hair, which he did not have, and stirring the bamboo wind-bells to their faint, hollow music. At home again I sat down and closed my eyes in serenity. At once I saw Miguel. He was, thank God, many miles away. He was way off there in the big valley of the Sacramento, or some small valley of this fruitful western country, picking apricots, picking pears, picking apples. And presently the figs would be ripe. As he threw his head back in that operatic gesture of combing his long hair, which he was never going to have, he opened his mouth and round fruit, full of sugars and juice, fell into it. Money fell from orchard boughs into his empty Mexican pockets. Far away in Michoacan the eyes of his hungry family grew bright. This clairvoyant scene gave me much reassurance. My conscience was at peace. California was full of ripening counties. Contra Costa, Sonoma, Lake, Napa, to begin with. Miguel would go from one to another, harvesting the rich fruits. I could stop conjugating the verb *ser*. Exalted, I wandered in my empty garden. All day it was very quiet, and I smiled at my husband who nodded his head at my contentment.

In the evening the telephone rang. I should have known better than to answer it. Sra. Mendoza spoke. "God bless you, honey, Miguel is here." The light of legend that had illuminated my day blew out suddenly. "He didn't go to the big valley?" I cried. "Oh, yes, honey, he went." "And what happened?" "Over there they say, for Chris' sake don't send us no more fruit-pickers. We got too many." "Oh, no, no!" "Well, see, honey, Miguel is here right now waiting for you tonight to get him at eight o'clock tomorrow morning." The disappointment was heavy as the slight, hand-made myth full of neat, voluptuous orchards and currency of the U.S.A. spinning down like leaves all collapsed. I made one aimless effort to revive it, at least to hold it in sight in an unworthy way. "Tell Miguel to bring tortillas for his luncheon." "Tortillas!" the senora's voice was full of honest amazement. "How do you mean tortillas, honey? He's got no money to buy tortillas." "Oh, I understand." So back again to the thick Cassell's. What is the Spanish for peanut butter? "Bye bye, honey, God bless you."

I went to bed early, in dejection. Poor Miguel. No money for tortillas. That would take the nightingale out of anybody's whistle. I had no work to offer him, but I must think of something while I slept. Perhaps to rake up the withering fruit that fell slowly through my dreams.

JACINTO

IN CALIFORNIA almost any country property is referred to as a "ranch." When we bought an old ranch in the Napa Valley we did not expect to become old ranchers, but ownership includes responsibility and of the latter there were soon many kinds. Suburban life had not prepared me for the one I found myself involved in when I became the caretaker of a goat. Our son had acquired a small flock to provide milk for his children and by informal steps and for ill-defined reasons I turned up, surprisingly, as the meek guardian of a determined billy goat whose name was Jacinto. I prefer to say that I was Jacinto's counselor. In this case guardian means goatherd and that was beyond my capability. "Counselor to a smart goat," that gave me a better position, at least in my own sight, and even suggested that the goat might be listening to me, although that was improbable. However, I found that a title, the only one I had ever had, and self-bestowed, was sustaining. It defined, for a summer month or two a few years ago, my hopeful attitude toward a conceited animal.

I had much to learn about goats. Jacinto had nothing to learn about me, since I am only human, a subject goats have studied for ages. What else had they to occupy their lively sensitivities and intense curiosity all those long afternoons in life and legend in the early centuries of their Persian masters, or waiting with Philoteus for Odysseus to return, or being worshipped in gold in the vineyards of Phlius, or eaten raw in the name of Dionysius, or filling the ancient woodlands with music and scandal of

Satyrs, Pans and Fauns? A goat is a deceptive and mysterious entity. He begins life as the most lovable and charming of animal young and even in his advancing mid-existence he still has those softly hanging ears of hearkening velvet that look so dog-like and ready to lift in friendship at your voice: but what of his eyes? Sardonic and calculating, agates and fathomless, kingly and proud. He has been both god and sacrifice, slave and feast, pet and symbol. His royalty is in his eyeball, and hilarious arrogance in the sudden grace with which he knocks you down.

As we start along the ranch road this nice morning of summer Jacinto stops, and consequently I stop also at the other end of his chain. I have not as yet figured out any way of continuing until he is of a mind to get in motion. It is like this every time. This is where counselorship assists me. If I try authority or even firm suggestion it is no good, and I am a failure. If mild advice is also no good then at least I have done something less drastic and the goat's refusal to accept it leaves both of us less antagonized. The relationship is psychologically fertile, and Jacinto makes the most of it as he turns to look at me with his pale severe eyes. He even takes a few steps forward. "Good and amiable goat," I exclaim. "And Jacinto," I add politely, "just get that branch of elderberry, the crooked one." He takes all the elderberry and moves a few steps. Then he stops again. This stratagem continues until an idea occurs to me. In this quiet, bright sauntering A.M. it is easy to start remembering. Remembering what? For me perhaps a moment of childhood brought back by the pungence of madia. But for the goat whose memory is racial and primordial, what memories he could be having! Uninhibited, undiluted, they flow over him like water. Like water over a stone. So it is necessary to stop and to stand still. And I stop with the goat and stand still with the goat. He is remembering other valleys, almost inhumanly ancient, grey and harsh, where he left his profile on the first walls and taught the people to celebrate; and later gracious valleys where men and women sang to sweeten the ripening grapes. He is remembering—wait a minute. Sheer fantasy, straight out of my own head. If that solid goat could speak I know what he would call me.

Jacinto was of the third generation of the small flock on our place. Of the group I had been best acquainted with Momma, a sweet friendly creature quite different from her son Jacinto. And intelligent, too. John fixed a cart for her to stand in that made it easier to milk her at a convenient level, and when she was ready to be milked she climbed into the cart and called for him to come. Or sometimes she would sit there resting, looking comfortable and comic. She enjoyed being petted and in return offered a few soft nibbles here and there. One afternoon our neighbors, June and John Harmon, who had come to do some mysterious water-witching for our well, tried to teach me how to milk Momma was who sitting in her cart. I was not good at the primitive thievery of milking, although June's advice was clear and spirited. "From port to starboard," she said. But it is not easy to pull milk, although once the maneuvre is begun it seems enjoyable to all concerned.

Poor Momma, I cannot think of her without overwhelming sympathy and also anger. I was alone here one evening; everyone else was gone somewhere. I heard barking, and also a strange kind of horn-like cry that I did not recognize. The sounds were coming from the knoll at the front of the place almost a block away. Later we learned that a vicious dog was attacking Momma and she was crying for help. I can never forgive myself for my dullness. I thought the odd horn-like cry was the sound of a child's toy in the distance. And so the predator continued his attack and the goat, tethered to her stake, had no defense except to whirl and plunge. Repeatedly the dog ripped her udder. Poor agonized terrified Momma, when we finally discovered what had happened we thought she would die out of her misery at once. But she did not, and when we called the veterinary he took the entire next day to arrive and was of no help except to give her the drug that brought the end. She died with her head in my lap and my tears pouring down on her. The dog, belonging to someone nearby, was identified, and John shot him, but the futility of revenge afforded us small satisfaction. When we hopefully came here to live in a part of the state new to us, we did not find it hard to begin to take root possessively in a landscape that was so appealing. But there are sad unchosen things that may crowd into a new image of home. We had not thought to encounter them yet here on another map in another latitude, we must write them down plain, the local and universal names of remorse and pain. Roots.

Before the poor nanny goat's tragic end, a half dozen sheep had been added to the goats. This roving community soon taught us some pastoral lessons. "What a charming little flock they make," we had said in ignorant pleasure. "What a romantic detail they add to the scene." When we turned to look again we found that they had eaten a part of the scene to which we were dearly attached, our treasure of wild flowers, at least three quarters of an acre of gold calochortus, the lovely mariposas lifted on tall stems in the grass, and also a rosy meadow of gillia, dainty examples of pointilism, so delightfully dotted and maculate as to make of each small flower a fetching creation. These native flowers had come as gifts to us from our own land. In the spring of their best display they were quickly wiped out. Ferns under the oak trees, too, all gone, and in the orchard the young prune trees were chopped into as far up the trunks as possible. By standing on their hind legs and climbing, the unstopping animals even destroyed the lower limbs of several fine cedars that now struggle to renew their torn branches. The velocity of hunger was incredible and began with taking the beautiful buckeye that grew between the lichen-covered rocks under the oaks. When all these, and other, acts of corruption were discovered by us we were quite speechless until my husband, a man of consummate restraint, broke into profane indignation. Our son, who likes plain uncovered earth, took it calmly, but agreed to some fences and coralling. Their united ambitions had led the small flock to consume everything within reach on top of the earth—our new appealing earth—omitting only a few unsavory stones, and they did the best they could about these when the goats found an old prune-dipping vat set with mortar into a frame of boulders. They delighted in taking the structure apart. It must have been standing where it was for fifty years. Its abrupt dislocation disturbed three stout elderly resident rattlesnakes. We made a point of presenting the rattlesnakes to our son, saying he could have *all* the bums.

But these recollections drag, and Jacinto's present poor manners bore me. There are many things I should be doing. If I try to push Jacinto he will push *me*, and much harder. Just then I saw my husband approaching along the ranch road on his way to the house. "Fred," I cried to him peevishly, "this goat has just called me a fool." Fred came to a stop and looked at me hard. Then he began to laugh. "He can't stand your hat." The hat had come to me by charity of a friend who got it in Acapulco, a baroque seduction of southern craft. It had bad-smelling little shells that dangled from the brim and it was far too large in the crown, so that no matter how straight I stood I seemed to crouch, and it was garishly decorated with varnished flowers that pleased me very much, being tropical epiphillums. Well, maybe it *was* a joke. It nodded and flopped in false sociability, for I thought of it as a place to hide. I was surprised that my husband had ever noticed it. Pure awfulness he could ignore, and it took chic or originality in feminine garb to draw his attention. "Ever noticed it?" he exclaimed, "How could I escape seeing it? That hat is one of the two worst hats ever beheld." "And the other one?" "The famous one in full sail painted by Matisse on the head of his wife." "And do you think Jacinto's opinion about anybody's hat is important?" My husband considered. As often at such a moment he used the "plain language" sometimes employed between us two. "Jacinto," he said, "when did I ever give thee leave to be so sassy?" The goat, who knew his name, shouted, "Bah!" It was my turn to laugh. But suddenly, "Oh, he's moving! Adiós, adiós! Thee can buy me another hat and I'll feed this one to thy friend here." I trotted after my charge crying, "If he's impertinent again I'll kick him," and Fred warned, "Thee must kick him carefully," then added sharply, "And thee can kick him once for me." Never will three quarters acre of our beautiful calochortus wantonly devoured be forgotten, even though Jacinto was not, unto the third generation in reverse, responsible himself for the crime.

The goat and I are headed for a ravine below the road. Jacinto and I both know the place well. It is shady and attractive and at first, several weeks ago, it was grassy, but now only the outer portions are grassy, for the goat has taken his favorite herbage, leaving just the stubs of poison oak and rabbit brush sticking up as he circled around his tether, a strong fence-pole regularly moved to give him fresh grazing. This sounds simple, bucolic, efficient. In his own thorough way this description applied to Jacinto, but not to me. In my efforts to keep him contented I had often taken him a snack of prune branches laden with ripening fruit, or freshly cut long grape stems that sprout out from the old abandoned rootstock, an apple stem or two as well and many cuttings of Japanese quince and plenty of cherry and pear, all in all quite a heap of minor branches, stems and twigs and enduring vegetable matter. The leaves, fruits and tender parts disappeared, but the tough remains lay scattered about, until repeatedly swept into a tangled pile by the goat's chain as he moved like a Maypole dancer around his post. Goats are appreciated as scavengers of poison oak, but prefer the leaves to the shrub and in pulling at the twigs may tear the root-system up and break the underground stems that are as full of the poison juices as the rest. During the first days of Jacinto's pasturing in the ravine I had been careless in handling these underground parts, not easy to distinguish from underground parts of other native plants. I myself was easy to distinguish when I broke out into a bright red rash from contact with their virulence. At that point I would have been wise to take Fred's advice, "Have nothing more to do with the goat and the ravine." But there was no other natural pasture left for him, and moreover, and more importantly, I felt driven to cope with his problems for they had become my own problems.

As we approached the ravine, Jacinto was pulling on his chain and plunged down the steep bank just after he had passed a large oak tree at the edge of the road, leaving me on the other side of the tree. Inadvertently I slid down myself, and then to correct this error—mine, of course—I scrambled up to the road and hastily slid down again, after the goat, and barely managed to secure his chain to the post before he crossed the dry stream bed and tried to rush up into the orchard beyond. Now I survey the immediate scene and decide that it is a mess. My fault, of course. Too many snacks. If mixed with chain they make an abominable fellowship of trash, a cooperation among many inextricable and objectionable items from all of which a woman with good sense would flee, and stay fled. Holding to a limb of the oak tree, I pull myself up the bank to the road and head back to the house for a rake and the ration of barley that Jacinto will soon be calling for. Incidentally, he was growing stout. The barley, as always, is a success, but the bamboo leaf-rake I have chosen because of a fondness for bamboo is not adequate to attack the trash. Jacinto is at peace with his barley, so I again struggle up the side of the ravine and start toward the house, which is a block away. En route I realize how ill-prepared I am to enter the arena of combined goat and poison oak. I must not repeat my earlier mistake of unprotected handling of poison oak, no matter how driven I feel to introduce unnatural order into a natural environment and avoid the havoc I could foresee if chain and trash became well mingled, and the chain thus so reduced in length that the goat could no longer reach any herbage. All my fault because of those generous snacks. And now I take time, before returning to the troubled scene, to don heavy slacks, socks, long sleeves, goggles and gloves. The treasure from Acapulco is still on my head. By the time I have found the steel rake and pruning shears I judge right for the job, I am hot and tired. Once again in the ravine the goat and I faced each other across my array of gear and tackle. He seemed to approve finally, after a calculating look.

At this point he was wound only half-way around his Maypole, and I released him and started to cut down a tall stub of rabbit brush which was certain to snag him. The pruning shears were not big enough for the job, and telling myself that I was not big enough for the job either, I scratched my way up the bank again and wearily plodded back to the house, hunting for a hatchet. On the return trip I filled my pockets with ripe prunes while trying to decide how many blocks or miles I had traversed in this long yo-yo and see-saw, that felt like ten, twenty, thirty, all for that goat. While I was absent Jacinto had been circling the pole and scraping up a mingled heap of rubbish and stones in which, by now, most of his chain was skillfully woven. Perhaps six feet at his end was free for activity, of which he took advantage. As I came within range he untied my shoe-strings, then he tried a few buttons on my shirt, giving my wrist a firm pinch. Then he lowered his head. I was not prepared for this, but I quickly tapped one of his horns with the hatchet and he reared up on his hind feet with a fine show of superior strength. He jumped into the air, flying sideways; he bent down and tossed the earth; he hooked the tines of the steel rake and dragged it as far as he could. He was excitable, theatric. Was this the moment to kick him carefully? No. I handed him a ripe prune. There was a lull. Seizing the chance, I began to hack at the troublesome stub, already catching his chain. Standing on top of a boulder by the stream bed Jacinto watched attentively and reached down to nip me as often as I bent to hack at the stub. I tapped his horns, and together, with dodging on my part, we got the stub out and I threw it across the stream bed into the orchard. Now could I reach the rake without being charged? There was a nasty bright light in the goat's eye. He was in gallant shape. Not daring to be too deliberate and give him the impression that I was timid, staggering, merely a mortal woman, I made a sally and did get the rake, but dropped the hatchet in my haste, and Jacinto turned his interest to that tool. Ah, how good this was: a clear field, the opponent distracted, and my own nerves a bit restored. Now to get those damn sticks out of the chain, gather up all the rubbish and make a pile, one, two, three piles, while I side-step Jacinto, cajoling him with sweet names and tossing ripe prunes at him in moments of crisis, I at last, beyond belief, got the mess altogether into the orchard where eventually rain, wind, sun and even the soft glimmer of the moon would attend in nature to the grand finalities of dissolution. I was close to desertion and Jacinto was beginning to dance again. Snatching my tools I threw them ahead of me up the side of the ravine toward the road. Using the barley bucket as a shield I dragged the heavy rake with the rest and crawled up the bank. I was entitled now to the proverbial long-drawn breath, and enjoyed it deeply, out of the goat's radius. I was modestly satisfied with what I had done. I had coped. Now, like Mrs. Porter and her daughter, I was on my way to a bath in soda water.

Suddenly a scornful bleat caused me to look back and down into the ravine. My rake had disloged a stub of poison oak, and it had remained, overlooked in the strenuous and excited affair of cleaning up the environment. The goat's chain, like a busy snake, had already wound around it and he was captive again. Am I to be handmaiden unto a goat all day? The stub rears and appears to be quite permanent. Around it goes Jacinto, and around a tree nearby, uniting all three, goat, stub and tree, in a trinity that defies me. At least I see the truth. I have not coped. To admit failure is a kind of refreshment, and under the stifling warmth of my hat—still with me—I feel a revelation of cool air. Leaving the clippers at the road—they looked adequate for the stub—I jogged crazily back to the house, dragging the other tools, and returned with a bucket full of water. In all the twitter and fuming of the morning I had forgotten this essential need. Embracing the bucket full of water and the long clippers, I slosh as discreetly as possible down the bank, and while Jacinto takes a good drink I cut the last cursed stub as far down into the hard earth as I can manage. The final problem is, where can the bucket stand to be in reach and not be pulled over by the chain? I collect some heavy stones and drop them into the water. Alas, mud! Removing them quickly I say firmly to the bucket, "Stay!" as to a well-trained dog. Then for the ultimate time, I feverishly hope, I haul myself up the bank and scramble onto the road. I swore to myself that the ravine had deepened and steepened under the continual weight of my conscience. "Never underestimate the geological power of a woman," I murmured. Well, the clippers are with me but I have lost a good glove. I dare to look back. Jacinto is tasting it. Should I give him my hat, with Fred's congratulations? Not until I have cut off the pretty salty shells to keep. And do not look back again, I said. But unexpectedly he was calling and it sounded important. Not bleating, but almost language. It startled me. It even sounded familiar—taunting, abrasive, lyrical. I could not relate it to anything else, and it ceased at once. Things float perilously in and out of one's head, especially in late summer, that random weather when stems are letting go and the tiny veins in leaves no longer function. I stole one more look down at Jacinto. He also looked up at me and we stared at each other with a considerable long stare. Although I was standing so far above him, his gaze at me was one of complete condescension. I had not expected gratitude for my foolish hard labors—(it was even difficult for me to say they were reasonable)—but neither had I anticipated being snubbed by a rowdy billy goat. I would not forget the expression in his timeless eyes. I would eventually perhaps feel flattered to recall that this haughty, transparent concentration of lens and light had been, and without stint, directed at me.

I hurried home, and in the garden, with a glad sigh of renunciation, I hung my hat on a large agave. As I went in the door Fred greeted me. "You appear to be in need of food and drink. First of all, drink." "It's the truth," I said, as I gratefully sank into a reliable kitchen chair, a support that did not require climbing up to nor sliding down upon. Just a good old hard unattractive chair I would some day throw away. Fred mopped my damp cheek. "Had enough of that goat?" he inquired sympathetically. "Definitely for today," I said. "But to begin with, I've made a discovery." Fred was always tantalizing. "An important discovery? World-changing?" I tried to sound deliberate and full of interesting thoughts. "Well, no. Just a bizarre discovery." "I like that kind," he responded, and was getting a bottle of Johannisberg Riesling from the refrigerator. I paused, hoping to stir his curiosity, but an idea that had furtively entered my mind as I walked to the house was ready to be communicated. I began with a flourish,

H: That goat of yours—by the way, I forgot to kick him.

F: Hurry back and do it.

H: That goat of yours—

F: Wait a minute. This begins badly. It sounds like a liability case. Am I protected? Anyway, the goat belongs to you.

H: No, he doesn't. He belongs to John.

F: So he does. Proceed.

H: That goat of our son John (slowly and distinctly) . . . Are you listening?

F: A little. (putting wine glasses in the freezer compartment to chill quickly) Go on.
H: (Positively) That goat. . .looks like. . .Ezra Pound.

For the second time in a few hours my husband was laughing at me heartily. Should I feel mortified? Condescended to? I decided against it and said,

H: You think the presence on our property of such a phenomenon is merely comical?
F: But I've always thought he looked like Ezra Pound.
H: Then why didn't you ever say so?
F: Well, it seemed just too splendidly disrespectful.
H: To which one?

Fred handed me a glass of Riesling. "Rejoice," he said. I accepted the wine happily, but without its help I was surprised by a sudden rush of loyalty. "You know very well," I said with dignity, "that Jacinto is a great goat." Fred looked exceedingly surprised himself and exasperated as well.

F: A great goat? What's so great about Jacinto? Let's get this straight. An unworthy, self-infatuated animal, and poorly arranged—
H: Whatever do you mean?
F: Too much in too many places.
H: Oh.
F: And always opinionated. AND full of tantrums.
H: You should have seen him dancing.
F: AND full of sin.
H: He enjoys it.
F: Without doubt he ate the calochortus.
H: (Shouting) He wasn't even begotten yet.
F: All right, all right. And when does the great Jacinto start writing Cantos?

The perfect question, the demolisher! Could I cope with that? I lifted my glass of beautiful cold glowing wine. “Any day,” I boasted, “just any day now.”

A BRIEF CHERISHING

FOR A LONG TIME I stand in my garden and look across the Napa Valley to the opposite hills. In this rich scene there must be a meaning that is special and about to become an image of this place, a legend of this earth. But it is finally with disappointment, although always with desire, that I turn from the view of vineyards and hills and come into my home. It might be that from inside and looking out, as through a lens, I will catch the haunting image for which I search, floating before the stony Mayacama Mountains. Could I trust its reality?

Then I recollect that in Southern California, where I lived for a long time, there was indeed a legend. There were portents and incredible events following. It was in part foolish and preposterous, even juvenile, put together from the euphoria of a famous environment and a divine climate, and rendering its originators too conceited to be tolerated, as if they had made the whole place themselves and were then resting on the seventh day. But there are those who have doted more sensibly and more humbly and who know the land much better, in drought, wind, mud, flood and fire.

We should take care how we talk about legends and that sort of thing. They are rife in a time when the old faith of men has been renounced, and the new certainties are terrifying, and it is easy to settle our belief on a destructive image. Ever beyond the vine and its fruit and the pouring of the wine, the heart still waits for a sign. May its meaning be gentle and familiar; above all, may its meaning be close to the earth. Wherever we are today, no matter how beautiful the valley, we are afraid to look into the sky for a sign. Let us look under our feet for a sign.

WHY HAD WE COME to live in the Napa Valley, my husband Frederick Monhoff and I?

The Los Angeles Basin and the San Gabriel Valley where we had lived with pleasure for nearly fifty years had become so changed by smog, freeways and expanding population that we had decided to search for a more reasonable locality. While we were in the midst of our search, my husband's profession, that of architecture, brought him to the Napa Valley on a commission for a client. I came with him. South is my favorite direction. At all times it has been my dearest destination. I came north with my husband five hundred miles on his professional errand. We were quickly captured by the freshness and variety of what we saw. It was very easy to like the Napa Valley, a place of knolls and pastures, or brooksides even along the public roads, and a look of rural security in uncontaminated air. We were not such believing fools as to think that any region in the west would remain permanently remote, or that a landscape that is appealing would appeal only to a chosen few and seal itself away. But this landscape did not look to be in danger. Pastures, vineyards, orchards and woods, enough of the land occupied for use and the rest held by nature in what seemed the sure embrace of safety. We decided to return and look farther in the valley.

After considerable searching, finding, excitement and rejection we discovered a piece of thirty acres, an old ranch with an old ranch house and remains of early vineyards and prune orchards. Fred first of all ascertained that the water situation was good for domestic springs with a well to be added for irrigation. He was not one to be swept off his feet by glamour, although the beauty of environment we found here was a strong influence, a necessary influence. It was an oddly shaped piece of land we had found, resulting from former divisions. The deed quaintly specified boundaries by chain-lengths, so many chains from here, a big rock, to there, an old stump. The first deed to the property was dated 1878, and was hand-written in letters of spidery lace. A long narrow thirty acres stretched back from the County Highway to Nash Creek Canyon. The road that we followed on foot back into the woods on our first exploration had been obliterated by the weedy growth of genista, called broom hereabout, a ravenous woody plant with a yellow pea blossom. It has consumed countless acres in this valley. We knew that a road was there because it was indicated on an old map. We broke through, and later a bulldozer cleared the way. We found giant firs and redwood circles, ferns, and a moderate slope down to the creek. It was true forest and lofty green peace. The towering trees at once moved Fred with their height and majesty. I also felt their power, but it was oak trees that I had longed to find, the evergreen oaks like those in the San Diego back country or, indeed those common to foothills of this valley. It was at that time that the spreading oak tree still drew me with a sense of familiarity and fraternity. The grandeur of the northern trees was alien. I had no memories in their high branches, no custom of returning to them for rest or shade. I saw at once that Fred had a superior feeling for these northern trees. His emotions were never divided by sentimentality, like mine, but were single and whole, being aesthetic or philosophic facts. How fine to have a mind like that, I often thought. And beyond question there was great attraction here for me, as well, in the soughing of the stirring boughs, and arboreal heights such as we had never owned. Could I learn to be at home at the feet of vast straight natives, trees whose heads were almost out of sight against the clear sky? I saw that the diminutive, domestic wren felt at home here. I saw that Fred, involved as he had been in long activity and the scenery of the south, could feel at home here. And the opening road led on down to Nash Creek where the small pools were clear and the cool water ran among mossy boulders and the salamanders lay without going or coming, as though they had never learned to move.

It was not long before we were the owners of the thirty acres with the tall trees and their verdant serenity. The year was 1953.

TEN YEARS PASSED before we were ready to move to northern California. The eventual ordeal of departure need not be included here. It is implicit in my reference in other pages to what was incalculably dear to us in the home and garden of thirty-six years in the south. Departure was a very long process, typical of people like ourselves who had never thrown anything away and were now under too much pressure to decide what to do without. It was more than quandary. We suffered, realizing how frail our judgment was, how little we dared rely on plain good sense. We had no plain good sense. Everything became irreplaceable, beloved, or worse, holy. Also it was unthinkable that we should leave our many kinds of plants behind. A few weedier, watery sorts we decided to abandon climbed in pots with better varieties and came along as stowaways. Repeatedly we filled the Willys station wagon with shrubs, bamboo and garden treasures, and often alone I would drive north, arriving in the dark of midnight to see the orchard here eerily illuminated by the glistening eyes of deer as my headlights reflected all through the trees. Small lanterns of abalone, as it were, hung from bough to bough where the deer were browsing.

I must often have been a spectacle as I brought my cargo north, sometimes with a trailer for extra burdens. A station attendant once asked me, after a long stare, "Are you coming or going?" It could have made no difference, but obviously he thought that a woman alone, hauling an aboretum, called for an explanation. Here at our new property we hired a man to put things in a nursery bed until we could plant them where they were to stay. Back and forth, back and forth, how many times we went, the two of us, or alone, or with our son, until 1962 when Fred retired from more than twenty years of teaching at the Otis Art Institute, and also retired from his work as Principal Architect of Design for Los Angeles County. He was now ready to take the ultimate step. We finally came north to stay. He designed and built our new home on a knoll overlooking the Napa Valley.

IN THE TEN YEARS since we purchased our property the economics of the valley had grown argumentative. On occasional evenings we attended meetings in the high school in St. Helena. These were meetings of contention and outright temper. We found that the area in which we had recently settled in 1962 was sharply divided by ambitions for property development on the one hand and, on the other, by rather modest efforts to conserve the agricultural establishment. We had made an unwelcome discovery. We could not say that it was the same situation we knew before, for in Southern California there had been no argument, no struggle to preserve the agricultural character of a vast citrus valley. The San Gabriel Valley had been wiped out with bizarre speed and ease. Now in the north we were shaken to hear vineyardists heatedly contending for the too familiar privilege to convert their rich and productive land into building lots and cash. Strangers though we were, we stood up boldly in these public meetings and solemnly, loudly told them that they did not know what they were doing. "You will lose the valley," we cried, "it will be one big subdivision." And we told them, "We come from Southern California," as if that were the same as saying that we were escapees from a fate that gave us a desperate vision. Fortunately, enough people realized that it was an historical moment for the Napa Valley. Out of the contention and the often repeated assertion that "his constitutional rights give a man the privilege to do what he likes with his own property" came the sanity of the Agricultural Preserve which still functions and saves good prime land for farm uses and ownership, and has kept the subdividers subdued.

My husband gave much of his time and energy, after we became settled, to the problems he had not expected to find in his new home but was ready to face. It was his way of becoming a citizen without delay. When I begged him, in his retirement, to turn again to the painting and etching of his younger years he said, "This is more important. It has community meaning." I was indignant and insisted, "Your art has community meaning." He thanked me but added, "Think how much room it takes on the walls." It was an argument that would never end between us.

WITH THE YEAR 1962 we had come to the date from which I myself was looking both ahead and back. Fred rarely looked back. He thought it a poor habit. And I knew that, no matter how much I was drawn to look back, I must look steadfastly ahead. Although no longer young, we had to make young choices as if the years ahead were meant for discoveries and fulfillments. We were not thinking of threats and ironies. We were together and could support each other, each other and our kin, for our son John, no longer in college, had come with us, and eventually Fred would remodel the old ranch house below the orchard for the occupancy of John and his family.

I have always envied people who have roots. I brought my envy with me. It would also seem that with a son, and later on four grandchidren here, I had brought my roots with me. But rootedness is a condition of continuing desire. The wistful want of roots is where roots begin. Is it at all likely that other people, reading these words, will have patience with my need to be at ease in a new place? Can it seem to others to hold the dilemma of difficult importance that it holds for me? Fred and I came to the problem in different ways. In his work in the south he had acquired social and ecological interests. In addition to a warm collaboration with his clients he gained a serious attitude toward a citizen's participation in local affairs. He knows what it is to work with the local bureaucracy. He is well-read in modern social literature and philosophy of the environment. He has ideas and can present them. As for me, I have been only a citizen in my emotions, expressive perhaps, but timid. Let's say I am at least present. I get out the wine glasses when Fred invites people to our home to discuss a current problem. I remind him that we can seat only twenty guests comfortably, including a few on the floor. I toast and salt a great many nuts. I admire my husband as he presents questions and suggests responses, to which I listen in silence. And I believe that, if I were not too inexperienced, there are two things I could talk about that have not been mentioned by our company, two things that might elevate me to participating citizenship, and that would either make me a good citizen or hard to get along with, two things: *anxiety* and *conscience*.

FOR A NUMBER OF years Fred had been going to southern Illinois on a very considerable architectural job for the Henricks family, publishers. These trips continued after we moved north. He was absent from home, as it happened, at a time when a catastrophe threatened.

Which day of the week? Which hour of the day? The month was September and the year 1964. It was one of the most vicious fires in all the years of the burning northern counties. It must have been on Sunday that we all knew, down on the valley floor, a fire was present up there near the Palisades, and visible above Nelda Cordy's house. I saw a fireman filling his tank in Calistoga. "Can't you get it out before it grows worse?" "Lady," he answered, "There's nothing up there. Only rocks and rattlesnakes." In other words, let it burn. And it did burn! Once really started it roared down the mountain of St. Helena toward the town. It came down in immense flaming cascades, mobile immensities of conflagration. Language is not meant to deal with such fire, and words can only catch fire themselves, inflammable in the sight of the ghastliness that descended cliffs and wooded slopes. It was five miles away when I first saw it but overwhelming, and coming on.

Fred was half a continent away, but John was in his house, and after the first horrified view of the hopeless scene on Monday night I went to the phone. "You see what I'm looking at?" "Yes. I have a good view. Awful." In silence we hung on to our appalled communication. We sat and stared at each other helplessly over the telephone. "Nothing we can do." John said finally. "Nothing we can do." I agreed, and reluctantly hung up.

Restlessly I got in my car and started up the highway to Calistoga. I was surprised to see the road lined on both sides with people and automobiles. The town was being evacuated, but I could not turn around. When I came to the intersection and tried to turn to my right, an officer stopped me. "You can't come down Lincoln into town," he said. "What do you want?" "Just to know what's happening." He spared a second to stare at me, then stopped the traffic streaming out of town and directed me to turn around. "Where do you live?" he asked. I motioned in the direction from which I had come. "Back there, near Diamond Mountain." "Honey," he commanded, "go home."

I went home and tried to get Fred by phone, but there was no connection at that hour of the middle west in his motel. I talked instead with the chief of police and left a message. By this time a strong wind was adding to the terror of the fire. I called Mr. Wright in Napa. He worked for us and had always shown an old-fashioned devotion to our affairs, remarkable in our time. He drove the thirty miles to contribute his goodwill, bringing his son with him. "If Eldon's weight hadn't been in the car I'd have been blown clean off the road," he declared. We all sat and stared out the front windows at the huge flames as they continued to rage on the mountain across the valley. I suppose that eventually someone lay down on the floor and went to sleep. I don't remember. Under the stress of that night and the next day time went by in a horrid dreamlike way. I pulled paintings and carvings out of the house, and into the Willys. I went down to the state park where the fire-fighters were camped, eating or sleeping exhausted on picnic tables in the midst of trucks, rigs and tankers. I saw the Navajo boys, and felt reassured by their presence. Known to be excellent fire-fighters they had been flown from Arizona or New Mexico. I came home and added books to the pile in the station wagon. I listened to sirens screaming up and down the valley. John came up and we talked. He was worried about the dry grass on the bank below his house. Fred called for the fourth time for news. At evening the smoke was so thick that I could not see to the bottom of the orchard, although there was moonlight. I could hear the deer moving away from the house when I went out on the deck. I could hear the click click of their hooves and I caught sight of two large bucks with antlers. Wilfong, the carpenter, who stayed close by in Fred's absence (people are wonderful, I kept saying), told me that all day, all night, the deer were crossing the valley from the Silverado side where it was burning. Several times I tried to add more things to the station wagon, and was finally deterred by Mr. Wright, who was still present. "Show a little judgment," he said firmly, in his clipped Nova Scotian English, and the carpenter added, "I built this house. It's well built. It can't burn." And it didn't burn! God bless it! But at some time on Tuesday John Nolasco's pasture over the highway was burning and the fire crept across to our side and started the roadside grass to blaze. Out of somewhere an angel by the name of Smitty jumped on the spot of fire. God Almighty gave Smitty a push and Smitty jumped hard on that fire, while all about the counties and the hills were blazing in the wind—Sonoma, Santa Rosa, Mount St. Helena, Pope Valley, Knight's Valley, Silverado side of Napa Valley, and back-firing below Angwin and the sanitarium. All these places in flames with others at once, an inferno. Our next neighbors, the Bosc's, hearing that Fred was gone, came over to see me and André said, "Turn your pumps on, dear, and fill your reservoirs." This was a generous and lordly incident in the midst of an argument over contiguous water rights. Calamity maketh the heart gallant.

These things and more I recall from those few terrible days in September. At that time we never travelled by plane. When Fred reached home by train the fire was out. Only two years after we had built our new home we now had something to say to each other about the obstinate, exalted and feeble-minded love of Californians for the land that destroys them. Forty years earlier my mother and I had lost our home and all we possessed to the Berkeley fire of 1923. Later in 1982 I was to witness the identical hideous descent, as in 1964, of mammoth flames down the slopes of Mount St. Helena. And every summer all over the state the news is redolent with the sweet fatal odor of burning brush. Why do we come? Why do we stay? We stay because, as we look around, beguiled and happy, the risks are out of sight, the risks are sunk deep in our conviction that, aside from occasional viciousness, it is a perfect place in which to live.

Moreover, there is always a fascination in the final and ultimate thing, and it draws us to possess it. As the foam of the Pacific washes over our feet, frantically facing west we can see no more of California lying ahead.

THE HOUSE HAD STOOD and faced the fire of 1964 with its windows full of a raging sight and its garden just beginning to get into the ground. The house, together with its originator and our family, was to become the most significant thing in my life. From where it lies on a knoll it views a long stretch of valley, vineyards and mountains that fill the memory of the beholder with all the weathers, seasons, skies and birds that northern California can offer. Yet it is not misplaced in an immensity, for trees and colors of rock and movement of light give variety and charm to a scene always appealing, never too large, and just wild enough. When Fred first chose the site for our house and started to build, the view was all native green of chaparral and fir. The scene was dominated by the great Palisades that lie along the ridge of the Mayacamas opposite, and we took a special bliss in the sight of their dark majesty. Before we had established ourselves, however, we faced a disappointment. There is a knoll in the middle of the valley facing our property. We watched in distress and bewilderment as huge mechanical equipment attacked the crown of this knoll and gradually reduced it to flat areas. On this flat ground slowly rose a mammoth building of a blank white, a new winery. Its enormous whiteness now dominates the scene as the Palisades dominated the scene originally. In the twenty years since it was built no tree—pardon me—one slim Italian cypress tree has been planted to mask its unyielding command of the hilltop. Fred felt so keenly about the intrusion of this insensitive construction that he invited one of the owners to view the effect from our windows, suggesting a more subtle coloring of the exterior. The gentleman accommodated him with a call. Leading his visitor to the window Fred pointed, with a deprecating gesture, and said dramatically, "There's your building." Our visitor threw up his arms in triumph. "Magnificent!" he cried. There the matter ended. To judge by the people who stop along the highway and take photographs there must be many tourists to whom a very large stark building, egregiously white and aloof to its surroundings, is magnificent. And a great many other people are confused as to its purpose. Staring from our windows they ask, "What is it? A hospital? A monastery? Oh! A prison!"

To return to our own hill and our home. It is an enchanting long house, only one room wide except where there is incorporated an area for tender plants, semi-circular and going down to a lower level, my husband's studio. The dwelling extends, invitingly, on two sides of an ample courtyard. Throughout it is a talented association of use and style. I must be pardoned for assuming that my love and devotion to the house are proper means for presenting its image. And I am sure that to describe it in detail is the wrong way to say that to live in this house is a constant and comfortable joy. It is complex and subtle, yet in all ways simple and convincing, never confusing. It is magically related to the surrounding view as house and environment open into each other. Between the living room and the adjacent plant area is a wall of glass which permits a company of graceful rhapis palms to be present in the living room itself while contained beind a transparent barrier. Neither indoors nor out do the flowing lines of the house become rectangular and geometric, although it is modern, made of this century. There are none of those cathedral ceilings, so often extolled, but in every room a ceiling strongly beamed and imaginatively tilted and rising to upper windows that show the sailing of the clouds, or the rising of the stars, or the rich flowers of the orange campsis vine when it blooms. And at night the reflections from much glass may turn even one globe into a mystery of marching lights that go on and on. Yet this house is modest, without pretension. It was built for use and pleasure. In its exterior tones of brown redwood and chaparral green and gray our house merges with the environment. Seen from a distance—as from the big white winery in the middle of our view—it virtually disappears: the right lack of domination as conceived by its creator.

Our hill has its share of local stones. They are excellent for building. Having long ago been fractured by the weight of passing centuries they now docilely fit together. They make a handsome house wall, and we have such a wall at either end of our house. It's a pleasant feeling to know that their strength and security come out of the earth under our feet. One day as Wilfong was building an extension of the west wall I said, "That's going to be extremely attractive, but I can't see through it." Fred was standing nearby and inquired, "Why do you want to see through it?" "Because over there in the woods is a fine oak, and just now it's a gold oak. I've been enjoying it as I work in the kitchen. Biscuits à la golden oak. Pretty soon, plain biscuits." Fred held up his hand, "Wait a minute, Mr. Wilfong. Have you ever built a stone wall with a large round window in it?" "I have not," said Wilfong emphatically and distastefully. "Want to try?" After a discouraging pause a gleam of adventure lit Wilfong's eye. In him stirred the blood of old Kentucky frontiersmen. At least, something stirred and he lit up. "If I can build a butterfly roof I can build a stone wall with a hole in it," said the surprising carpenter who had already turned into a fire-fighter and now into a stone mason. Thoughtfully and slowly he made a circular wooden frame, and by rotating it as he worked with the stones he got them in place and still achieved the opening.

Since then I have watched in delight a procession of many seasons of green, gold, russet, followed by scarlet of pistachio that stands beside the ivory trunk of a silver-leaved eucalyptus—all seen through a stone window. Are things different when seen through a stone window? Ah, yes! Different and more beautiful and exquisitely remembered.

WHEN WE OPEN the door after a heavy winter rain we can hear the running, dashing and shouting of water in the ravine to the west. When we open another door we can hear another brook in another ravine to the east, as it makes delightful Niagaras falling speedily between mossy rocks and filling clear pools and although it is a discreet size it creates an uproar. Almost overnight the winter ferns spring up along the water-courses, the banks above turn bright green and the hound's tongue starts to bloom with its vivid blue sparkle. As winter transforms itself into spring the first flower we look for is this one, standing up among its large, pointed leaves. We never pick the flowers but permit them to scatter their nutlet seeds. Another flower we look for is "the invisible one," the brown dappled fritillary that swings its bells in stony ground and at first is not there, but then its faint colors begin to beckon and it stands, to our surprise, where it has stood all along.

In the damp of winter and rainy spring many of the deciduous trees are hung and flaked over with pale green lichens and fast-growing moss. It is a metallic green. The mosses break into puffs and tiny branchings and, closely observed, into thready forms and soft clots of vegetal composition. The suddenness with which this transformation has occurred is mysterious. I ponder it. We knew that leaves were falling, we saw the oaks turn bare, but when, on what witch of a night did the trees stand and let the coppery green of winter take over? Winter, while the sasanquas are still blooming head-on in masses of pink and white elegance, shedding their violet-like fragrance. Yet all about, the season and the weather produce illusions of a universe done up in green snow.

And I said to Fred, "Look at them, thousands of trees, all verdigree, but beautiful." "What do you mean?" he asked. And I said, "The whole valley is one obsession of moss and lichen." Fred looked at me carefully, then he said, "Come to the window." I went with him to the window in the dining room. "Show me." he demanded. I stared below and out and across. I saw the usual well-clad firs, ever-green oaks, pines and our own eucalypti. Where were the scores, the hundreds of moss and lichen-clad copper green trees? I was dumbfounded. "What happened to all of them?" I asked. Fred laughed. "Are you sure you saw them?" "Then what did I see?" "Just enough to convince you. But especially you've been looking at our own mirabelle right here in front of the kitchen window. I've seen you standing and staring at it. Our own mirabelle. It has really got hold of your imagination. You've been enjoying it without knowing that you were caught." "You're right," I answered, "and as usual I don't mind that too much. I've been wondering how many years ago it was that someone tried to turn this very tree into a useful prune tree but the mirabelle stock resisted and used her roots for herself. No good prunes for us out of her. And then you came along and built a house just above her, and you trimmed and shaped her branches and made a tall, spreading graceful garden tree out of a dumpy one." "Thank you," said Fred, "I accept the compliment." "But what about it? You made a rascal out of an innocent tree," I protested, "a she rascal." Fred tried to look astounded, as I continued. "A talented, green-haired sorceress," I insisted. "And she made a fool out of your wife."

A DISTURBING LOCAL EVENT we attended in our early days in the Napa Valley was the conference staged in Calistoga by the Department of Economics of the University of California at Berkeley. Let me recount this event as well as I can remember its details and its spirit. It was professionally presented and with a correctness that was to us highly threatening. These very educated men considered the Napa Valley to be a laboratory for the exhibition of social change from rural conditions to urban. Their conclusions were based on the incontrovertible facts of what had happened in other valleys, and bolstered by a mean swarm of statistics. The first speaker, however, compulsively described the beauty of the early morning scene as he had approached the upper end of the valley, and regretfully added his personal elegy at the thought that such delights might be doomed. Then came, by individual members of the department, a succession of lectures that left the environment divided and dissolved and remade according to the strict visions of the visiting scientists. In calculations of the urban future little remained of the rural dream. The popular country hope that was extending its promise to people from the cities received a serious rebuke at this well-planned conference. The men from the university were working hard to present their ideas persuasively, even moderately, but it was not an audience easy to persuade. Indeed, most of those present felt that to surrender rural and get urban conditions in return was a bleak exchange of value.

We noticed in the audience an elderly man whom we had met and knew slightly. He appeared to be upset. He owned a house in the country with a surpassing view of Mount St. Helena. He had been recently widowed. The image of his lost wife stood and held out her arms where the men from the university were just now cutting up a serene refuge and laying down blacktop. The elderly man and his wife had picked their choice property for the peace and pleasures of their last days. In his loss of her all that remained to him was the sanctity of place. It was being invaded and shattered. Suddenly, he stood up, interrupting the climax of some fine conclusion. "And what about me? What's to become of me?" he demanded in a quavering voice. There was a long, strained silence, and at last no answer at all from the scholars. They were slow to guess that they were stepping on tender hopes and decent visions. They must, to begin with, have been convinced that their own educated ideas were important far beyond simple attachment to sentiments of home and land. They must have discarded thoughtful sympathies for people they supposed about to lose their right of way across an emotional view of earth and hills, an ownership too ephemeral to matter to the professional assessor of values. They were not thinking how earth is ancient, holy, beautiful and hostile, and in its countless meanings too elusive to be dryly categorized. Instead, they called a recess and announced luncheon, a very good catered meal, at our expense, naturally. There was nothing haphazard about the conference. With the exception of the quavering question it all went smoothly, as if rehearsed.

As we drove home we were both at first sober and quiet. We did not want to admit that we were troubled. "The chicken was delicious," I admitted, "but that congregation of intellectuals depresses me." Fred tried to be fair. "They did their best. They have a message." But I contended, "They have a theory." "True," he conceded. And I hurried on, "A message worth delivering would have been a warning, a plea to the people of the valley to save the valley from the city." "But," said Fred, "the function of the intellectual is to define, clarify, prove. What a good time they were having." "I wasn't having a good time," I objected. "But," continued Fred, "everything they said was true, as they see the truth." "Rats!" "It was true, and quite possible, and without feeling." he declared. "That's it!" I answered hotly, "Without feeling." "But something equal to feeling," Fred went on, "is their conviction of perfectibility. They believe this grand spread of country land will become suburbs. They want to see it the best of suburbs." I waited for a frozen moment, then, "Do you expect that to happen?" "No. But if I listened to them long enough the suburbs would be climbing up my neck." I shuddered. "Cheer up," said Fred, "there's nothing more rural than fried chicken. We're saved. If they had served shrimp we would be doomed." At this moment we turned into the road that led up the ravine to our house. In the shade of the oaks the ferns were green and fresh, and the stones along the bank were still bright with winter moss. "Doesn't look like the city, does it?" said Fred. I laughed in agreement, and as we drove higher and came out from under the trees the sky above us was our own sky, and farther up and on, the hill was our own hill, and it would never be a city hill.

I AM RECOLLECTING that Fred said, "I did not come here to compete with younger architects." Nevertheless he was constantly occupied with plans and work at his drafting table downstairs. A number of private commissions came his way, and always he is interested in projects of benefit to a small town. He had dreamed for years of a community center to house meetings and assemblies of every kind for all citizens. After some urging by sincere sympathizers the city of Calistoga gave him $10,000 to build a community center. He said to me in disappointment, "For that amount I can build only a barn." With characteristic energy and determination he gathered volunteers, workers and builders of every craft, and found generous contributors of the necessary materials, and at last, stretching the city's meagre investment to fit the needs, he and his coworkers built a structure of one large room which would one day open through glass doors into a garden. On the day of dedication I sat proudly beside my husband. I knew that if the occasion gave him no professional pride it did move him to something he could prize quite as much—a warm gratitude to all the willing and devoted workers who had made his dream their own enthusiastic and hard-hammering vision. When a committee chairman presented the building to the city the Mayor rose and said, "I accept this building for the City of Calistoga." Then the Mayor sat down. With no further ceremony the years of use began.

Because he has had sympathy for those who cannot afford land and homes as costs increase, he began to approve of mobile homes and made a long and careful study of all the best within reach. To his own surprise he now says that it is possible to make an extremely attractive place of a mobile home park, with good landscaping, waterways and pools, shade and flowers, recreation and inviting club-houses. And twice, for clients with serious intentions he prepared elaborate drawings for such undertakings. Both times some kind of economic catastrophe overtook his clients and both schemes collapsed. The designs remain. I have enjoyed looking at them. The decorative patterns of many dwelling units and cars arranged skillfully in an artful landscape have the picture-appeal of ancient map-making, even the colors and novelties of a strange country sighed over by a wanderer in search of El Dorado. When I ask the architect whether my impression is true he laughs and says that he fears idleness and frivolity are sins that would flourish if I were to be appointed gate-keeper to this surprising park and its pavilions. "Just keep out," he begged, "unless you believe that life is real and life is earnest."

As for his art, the walls of our house are hung with many examples of his mature work. Among them are decorations for poems. I take a design into my hands and look and look at the clear engraving cut into boxwood, or the image, sure and lucid, black and white: rain drops falling into a pool, the bird soaring up into the song, the young girl drawing back in fright from the revelation of her own prayer leaning above her. Lyrically and sensitively and without any dazzlement of distraction the meaning of a poem is sensed and communicated, beyond the communication of language itself. Within these sure strokes of black and white the poem knows what it tried to say.

And years ago, when he was still teaching design to students he was teaching a new kind of design to himself, powerfully and inventively, in a series of extraordinary watercolor abstractions. They are inspired and solitary achievements of his talent. Neither before nor since had he created such mysterious and beautiful evocations of poise and mobility, eerily seized from nature or the hypnotic enigmas of modern devices. They are of this century but escape with strong grace from any hint of the geometric or intellectual styles. The control they manifest is of the emotions and the soul, yet they are not mystical nor evasive. Some of them are radiant; they could not be more radiant if they were painted with gems abounding. There are twenty of them and they are all mine, because Fred has given them to me, knowing my profound feeling for them.

One evening we were sitting on the deck in the summer darkness. We were waiting for meteors to race across the sky, as it was that time of the celestial calendar when the atmosphere of earth would be scraped and illuminated by sparks of glowing immensity rushing high overhead. We were saying how only one of our senses was involved in this viewing. A falling star, to the human spectator, is silent. Does it make a howl of turbulence as it drops through space? We do not know. A falling star, at that distance, has no odor, no sulphur, for smell or taste. It does not, thank God, touch us. Only with our eyes do we recognize, and for a shuddering instant, the glittering haste of heavenly trash sweeping through the universe. Patience is required to wait on the chance and magnificence of the solar discards; patience and concentration and revolving attention, together with warm pressure of a loving hand as the night grows cooler.

We supposed that we were alone with each other and the enormous night and its eternal spangling planets. But suddenly there was a cry from the bushes, the trees, the earth below, a cry of terror and supplication that rose from one passionate source and fell from everywhere. It was small and rasping. It hit us with the magnitude of a roar, and the tranquil hour was torn apart as we sat hanging in the shreds of night. Somewhere near at hand a lowly creature was begging for compassion. Somewhere near at hand a predator, in quiet fury, was tearing and biting through a soft furry throat while the victim, perhaps a voiceless rabbit screeched in agony big enough for an elephant and died too slowly. We had thought we were alone with everlasting stars and their abiding place. Now we knew the other great eternal one was present. In a few moments death finished what he was doing and, never to be filled, set off down the trembling hill into the darkness.

WHEN FRED CHOSE for our house the knoll on the center of our property there was nothing growing there except remnants of an old vineyard and orchard. These left-overs were removed. He designed a garden layout and I helped fill it in. It is possible for a gardener to be consumed by visions too strong for experience. The original impulse springs from the messianic writing in garden catalogues. After a year or more of hope and faith in commonplace perfection my pastel hollyhock seed turned out to have coal-black flowers that did not wilt in a decent way but dissolved in peculiar floral slush, and the sturdy clumps of famous perennials that made ladies so happy in past generations, were just corpulent daisies suffering from overweight. But still there is the distinguished umbrella pine and the handsome scratchy China fir, and the superb blue Atlas cedar, which was a gift from Fred one Christmas. These, together with eucalyptus of white and olive bark and rosy flowers, island ironwood and many bamboos and various magnolias have transformed the bare knoll to the point where Fred cries, "You're making a Rousseau memorial. Watch out for the black panther."

There is an easy-going formal garden with the beds edged by winebottles. Unlike the former redwood strips, they do not decay. In this valley wine bottles may be had at the back door of any restaurant. We were led to this solution by a book of old Chinese gardens in which we found garden beds in the shape of fans and vases all edged by wine bottles. We use chiefly a smaller size than the Oriental ones, the size formerly seen here on public dinner tables, less than a fifth. With any reasonable care they are rarely broken. They are appropriate in the Napa Valley. The flower beds they enclose are full of petunias in the summer and the bottles scarcely show. Even so, we have to agree with many wagers that it must have taken a long time and bottomless thirst to drink that much wine (not knowing who drank it).

Being from the south of the state we have been given to agaves, and I marvel now that we brought so many different kinds, all with murder in their sharp tips, yet no one has ever been impaled. A row of the blue kind *(Agave Mexicana)* bloomed a few years ago and in a latitude so far from their origin it was an imposing tropical memento. Also, some of our agaves are giants, and the giants are beginning to defy us.

I CAN CLAIM the terrace as my own idea. It is full of hellebores, ferns, anemones, day lilies, black bamboo and seedling oaks, each one of the latter a taunting problem—must I take it out?

When the terrace garden first emerged it was an empty slope of hillside occupied by a man with a bulldozer. As he stood beside his machine I said to Fred, "Couldn't he make me a terrace just here?" And my husband indulgently said "Yes." And for years I have been supplying that terrace garden with meaning. The best meaning has been supplied by Fred himself, however, in two finely designed benches that invite one to sit in peace and the sound of birds and contemplate with reverence the great blue oak *(Quercus Douglasianna)* that extends its boughs within reach. My own contribution, besides plants and bulbs, is a dragon. He is made of weathered floodwood and twisted forest wood. He is deeply satisfying to me, although he has never elicited any cries of praise from other people. He lies on a heap of red Konocti rock, stretched out regally (this is what I say) and surrounded by smooth, round water-worn stones. I gathered the stones at the beach, in fact I stole them, not knowing that it was a State Park. The stones are now bubbles in which the dragon is basking. This clears me. It is impossible to steal a bubble. The dragon has a haughty snout and a rippling tail. Not many women possess a dragon, even one bred at home or conceived in theft. He goes to my head. When his day lilies bloom I desire to sit down in the red stones and stay with him. When I ask Fred whether he likes my dragon he says, "I admire his attitude of aloof, antique wisdom. Is he T'ang?"

WE DID NOT part from a home of many years and many affections merely for the sake of casual ownership in a new place. Moreover, to live in the new place has been to encounter some of the same problems we had just turned away from. We had discovered that discouragement quite early. Here, too, the natural aspect is not safe. Here, too, the original is threatened. But wait—threatened by what? By the pressure of handsome vineyards. Should that disturb anybody? No, it should not, and it would not disturb us were we not apprehensively earth-aware people. However, the facts of success in an agricultural valley contain the seeds of danger. As new money pours in and new vineyards are planted, it is well known that much of the wealth being put into the ground is money from the outside, "foreign money," even exotic, Chinese or Arabian or Hawaiian. If over-planting of vineyards or overproduction of wine or disasters of competition strike the market it is obvious that in the suffering of money the land would be threatened, and quickly sacrificed. The outside investor has no roots here, no past. His sole local emotion is to succeed in his investment, to make money from what money he has committed to the earth. It is a nervous attachment. If his money sinks out of sight in an unprofitable vineyard he will retrieve it through subdivision if possible. In any case, the merciless charities of subdivision are all that the earth could depend on.

With others Fred has fought to defeat the freeway that was aimed straight up the middle of the valley. At one moment an actual threat, it has now been driven back into the drawing boards of the State, invisible and, everyone hopes, whipped and cringing. Such an episode makes for conviction that in any desirable place problems flourish because the eye of better business has brightened odiously at the sight of beautiful land. Where all is tempting and defenseless it is up to those who revere the earth to be strong. It is up to them to support faith in an earth that has not only economic value but the old values of regional character. The closer our feelings bring us to the landscape the more uncertain is our aesthetic ownership because of changes beyond our command. A stone bridge and oak trees are removed to straighten the road to town. An attractive country scene is lost. Who has affection for a mere utility highway, all grace and old oaks gone? And looking down into the valley one day we see another change. The river, formerly hidden by trees, is revealed, appearing as a plain pewter-colored canal. All verdure, above and alongside, has been stripped away. Flood control, we are told. Does it control, or let loose? And what of ferns, shade, moss, wildlife? What of the obsidian points and Indian artifacts swept down by winter freshets from vanished settlements of the Wappo higher up? The richness of the present crumbles away. The richness of the past washes downstream. These concerns come to us as we stand, hand in hand, at the window and look out over the valley. Because we both believe in these troubling conclusions they are at once worse and (touching hands) easier to face.

Unexpectedly, and in the mailbox, we find an acquaintance, even a friend. Someone has listed us as members of SAVE, and left the card for us to discover. Save America's Valleys From Extinction is an infant organization with an authoritative name. It promises "opposition to construction of a dam and reservoir complex in Knight's and Franz Valleys." We are pleased at the expression of a high-minded agitation in these adjacent valleys, and all for one dollar. I recognize the name of the secretary signed on the card as a collector of wild iris. A dam in Franz Valley would wipe them all out. Roots! Imperilled roots of a fastidious flower, a native aristocrat and, along with the oak-woods and romantic scene of its habitat, easy to be annihilated by human covetousness.

That which endures is safe only because it is out of reach. The High Rocks, the Palisades, the home of storms and rattling vipers, the resting place of hugeness and extremity, those enormous black cliffs, they are safe up there in their stunning proportions. They cannot be inhabited, they cannot be divided and sold, they cannot be washed away. If anyone dare to live there he must crouch at their hazardous feet, frightened by the grandeur and folly of his own courage.

SUMMER. At this point in time forget about time. It is too hot today to think of time, old rasping locust fit only to climb forever from one split second into the next. We are half asleep. The grapes are beginning to ripen. The starlings are in the vines and the mice enjoy the fallen plums. The pears are spoiling in the pails, and over to the woods our nuts are running away with long silver tails. Closing the gate never kept a squirrel from coming in. At this hot bright point in time we are half asleep and it is good to be hot and tired. All the sugars in vine and orchard are dizzy with calories and the scent of sweetness ferments in our nostrils. The bells at the winery on the hill are making a mellow hot sound, as if the hill itself were a great hot ringing gong. We are more than half asleep, carefully asleep, lest we fall from each other's arms into time's transparent tedium, lest we drift apart and sink, knocking into warm stones and burning spines and the pale scaffolding of yellow weeds.

IN ALL THE western seasons there is ambiguity. They mean one thing and commit themselves to another. They lie compounded. They are all mixed up. Autumn, for instance.

Spare and bright between two stones, hot stones,
I see the wild white oats the wind has left.
They vibrate and stand up,
Sunlight, moonlight have not struck them down bereft,
Long empty of their sap.
I see the sprightly skeleton of summer
Stand and not succumb,
And in this radiant small thing
Appear the grand intentions of the western earth

As up the tiny pipe of hollow stem
Something tremendous rises and is at work.
I do not know its name,
Not winter, not fall,
A northern season slowly out of its mind
And bowing into the hoot of southern wind,
Whose liquid bells nocturnal, bell to bell,
Swing strangely from the Border north
As up from Mexico sings the tropical rain.

Soon this sumptuous valley, yellow with old drought,
Will be renewed in luxuries of green
And northern forests nursing on
The black breast of a southern night.
But first permit the sleeper on the hill
To wake in joy and stare
Upon the fine dry bones of summer.
Most beautiful when plucked bare.

THE BOOKSHELVES of our house contain many volumes of philosophy, classical and contemporary, also psychology, psychiatry, and a mixture of mystics and sensitives and other fascinators. There is a great deal here, from the mighty Greeks to Ouspensky and the provoking Gurdjief, from the clarity of Athens and the wisdom of the Sufis to the relaxed gathering and incoherence of the latihan. What principles move what kind of mentality? Fred's curiosity is fathomless and his capacity for reading has no end. He has read them all. Since his disposition is what has been called "Taoistic"—having no desire to influence another person—he buys his books, devours them and does not urge me to read them, unless it be those of the English writer,

John Benett, whom he admires as a remarkable eccentric and varied human being, philosopher, idealist, educator, uncanny mathematician, gifted coal-miner, linguist and political leader.

At this point I recall a good story that I like to tell. It is short and full of truth. A visiting Buddhist priest came to call on us. I suspected that he hoped to promote our assistance in gathering a group of people to attend a course in meditation, to be held in our home. I was silently and earnestly resisting this idea. During conversation he suddenly turned to Fred. "Respected sir," he said, "you have a very fine aura." Then with a stiff bow to me, "The lady, your wife, not so good."

And on a rainy day, perhaps pathetically, and remembering my imperfect aura, I say to Fred, "Shall we dust your shelves of philosophy today?" Fred says briskly, "No, thank you." But I persist, because it is one of those mornings that seem lost even as they begin, damp and evasive. "Shall we dust your shelves of art today?" Fred looks at me fondly. "Certainly not." he says. "Why not?" I ask. "Because in my estimation the only way to dust a book is to read it." "Well," I answer, "considering how many books we have in this house that would be a long, slow dusting." "Just let the dust rest," he says. "Think how it gets badgered and blown around by broom and plow. It must be tired out." "There is something called The Good Housekeeping Institute," I responded. "You'll scandalize the ladies. You'll have them all fainting away." "A good idea," said Fred. "Have you ever thought," I continued, "how many feet of books we have in this house?" "That's a unique way to assess the frightening mind of man and his corruptible soul," he said, and I suggest, "Shan't I get the three-foot measuring stick and see?" Animated by what seemed to me a diversion on a dripping day I got the measuring stick and commenced. As I added up columns of threes and multiples of threes I remembered that there were books I hadn't been even thinking of—technical and art books downstairs in

Fred's studio, books behind books, cookbooks out of sight in the kitchen cupboard, privately printed books in rare editions behind glass, books in the bathrooms, books aloft on high shelves completely out of reach, more books stowed in dark small closets, and two book cases in the bedroom hall of botanical, horticultural and plant exploration books, besides piles of books about bamboo, all in addition to those in two front and one long back hallway. This list is not recited as a boast. It is a chant of desperation. "Perhaps this is a mad enterprise." I admitted uneasily. "Let's just pack up a big hamper of sandwiches and sit down and start reading with a sandwich in the other hand." "And where will you begin?" asked Fred. I said, "With the poetry of the Bible and Homer's Odyssey. Where will you begin?" "Jung's 'Dreams, Memories and Reminiscences' and 'The Owl and the Pussy Cat'." "Very good choice," I agreed, enviously. "And what sandwich? Peanut butter and jelly?" "Don't you dare!" shouted Fred, "I'll have the best caviar on thin dry rye toast."

As we sit at breakfast we can look out past the mass of the downhanging Banksia rose. If it is a bright morning following mist or rain the rose vine is hung with liquid gems as the sun catches a drop here and there and turns it into a sparkling diamond. But diamonds are too common. Sprinkled among them are other jewels, and it is these that cause us to cry out with surprise and delight until we sound like children and are, indeed, children, shedding years into a single drop of pure water. "There's a ruby!" and "There's a sapphire!" "Oh, look, an emerald!" And finally there is a rare brown topaz worth at least a little kingdom. As long as the sun is shining at the right angle the blazing drops hang, tremble and turn to wild, rapturous jewels, and we exclaim and pull each other to the best view. Strange, we never see an amethyst, but we almost do.

And while we are waiting for an amethyst Fred is apt to say, "Get out your old diaries. Let's read the old diaries." And I say, "Yes. Yes, I will." And for some strange, shy reason I never bring the old diaries to him. Why do I fail to do this? He would be diverted and bemused to go again on all our trips to deserts, mountains and spring valleys. He would remember with me the smell of the eucalyptus in the summer evening at Rincon, and the sound of the ducks rising from Laguna Hanson, making scores of small fountains as they took off and pulled the water up with their feet. And he would remember the oblique glance of the departing coyote near the cliff above Owens Valley where we were hunting for petroglyphs. We could relive it all together. Why did I hesitate to open those and many other pages now after so many years had passed? In truth I longed to show him those early records of intimate hours we spent together, perhaps on a walk in the Arroyo Seco or Millard's Canyon, alone and far from any thought that we owed responsibility or an anxious moment to anyone at home. Our freedom was tranquil and luxurious. Here in the canyon, under the laurel trees, this was the perfect and most dulcet place to stay forever. Here, in the canyon under the laurel trees, to desire and give pleasure, this was the only thing to do forever. Why was I too shy to open those early pages? Why does a human being throw away the chance of an hour of happiness whose memory holds the pledge of young love and the promise that is halcyon still, while on the sparkling vine the gems are shaking and the ruby slowly turns to amethyst?

ONE YEAR, after Christmas, we had a fall of snow. It did not melt as it usually does, but lay around looking pretty. From this valley on out toward the coast the woods and orchards were loaded with a weight of whiteness that was heavier than it looked, and caused many broken tree limbs and crushed chaparral throughout the area and made concern about summer fires. With our grandchildren in tow, Scott, Danny and Kelly (Lisa was a baby) we went up into our own woods to see how the snow storm was sitting. Soon a good deal of it was snow balls flying about and sitting in somebody's cold neck. All of this was highly exhilarating. Fred and I were inclined to enjoy the whiteness and silence of the woods with decorum, but that was no fun for the rest of our party.

As we walked along I took Kelly's hand and rubbed her cold fingers. "You got a fine little set of cake pans for your dolls at Christmas, didn't you?" I asked. "Oh, yes," she answered with enthusiasm. "Now you should make a batch of little cakes," I suggested. "That's just what I'm going to do the minute I get home," said the child.

We climbed on up the rocky road. The sharp stones were well concealed beneath the soft spread of snow, but walking was not easy. Here and there a bush of toyon still bore its berries of brilliant red, and along the banks the ferns persisted in their customary grace. The children howled and threw their snow balls and the boys were becoming more and more accurate. The old woods seemed to take pleasure in the young rioters in its white sanctuary.

Suddenly there was an ominous cracking sound somewhere ahead of us, and a crash. Fred halted the procession. "Sorry, chickadees," he called out. "No farther. It might not be safe. Something really came down that time." The children begged to go on. "Yes, I know, you hoodlums like to see things falling and coming to pieces, but no, not coming to pieces on top of you."

So we turned back, and the children were soon consoled of their disappointment by seeing me fall flat on the stony road. "Look at granma!" they screamed, "She's funny," and when I fell the second and the third time, their enjoyment, I thought, was really a mite excessive. Then Fred fell, perhaps overdoing the act with arms and legs waving about, while the merciless glee of childhood rang out strong and clear. I had begun to feel like that actress in the old movies (whose name escapes me) who was always stumbling over nothing at all, only she made an art and a career and doubtless a fortune out of it, but I remained merely awkward and helpless. As for the children, they dumped themselves on the snowy ground, jumping up to dump again, as if that were the natural way to travel.

So we slid and slipped down the hill and on home, coming to a stop at a pot of hot cocoa and a plate of cookies.

As I sat sipping my warm drink beside the living room fireplace, I thought about the ancient people of the valley, the Wappo. What did they do for comfort in the wet chill winter? Well, lacking a supply of Baker's chocolate, just about what I was doing. I remembered a circle of stones we had once found, deciding it to be a primitive hearth, and there they sat, stretching out their cold hands to the warmth.

THE TOWN HAS a city park bounded on one side by a creek and a few ducks. As a member of the Parks and Recreation Commission Fred is interested in the park. He is particularly concerned that tables and benches he is installing are not going to be tossed into the creek. "That would bother the ducks." he says. Is there a community in which senseless vandalism does not lift its boorish head? Bolts and concrete are the answer. "The pleasures of public conscience must be enjoyed by the few," says Fred, "and it is a lukewarm pleasure." I have helped him by writing letters that bring catalogues of boltable-unto-eternity park furniture. Now I empty the bolts from their muslin sacks, for the concrete slabs have been laid and heavy tables and benches obtained and they are waiting to be put in place, together with small barbecues. This is a good, useful moment. People can sit under the very fine evergreen oaks that line the creek and hear other people and children making merry. People can warm their beans and chili on the bolted barbecues and put the garbage in the trash cans that are chained to steel posts. A town well run. Friendly and welcoming. And trusting.

We have also planted trees for a children's playground, and the children who live across the street have had the opportunity to come over and knock them down. So far they have not run off with the sand box, and the swing is unwieldy to wrestle with. But we believe in them. We believe in their powers. "Now, now, no sarcasm, Señora," says Fred.

And one night, very late, we were sitting together in the park, alone, with the ducks on the bank nearby. It was almost midnight. There had been vandalism in the tool shed and Fred, ever the citizen, had volunteered to keep watch. On a summer night we sat and waited for the vandals. Another kind of thief, stealer of the peace, was present. I had heard of this one for some months, off and on, and knew that for the first time in his life my husband was struggling with a state of psychological depression.

"Why do you feel so depressed?" I asked. "I know it is a superior, spiritual trouble, but what is it?" "I can't quite tell you," he answered. "I suppose it is a condition of being alive today. I'm aware of a vast darkness and a sickening kind of helplessness." "How awful." I tried to realize the crisis he was caught in. "Is it the nuclear threat?" "Possibly. Partly. Science has created terrible facts that have brought moral responsibilities beyond what society or the individual can face." I thought for a moment, and said, "But you are not alone. Millions of people rebel against the merciless gifts of science. Millions of people come together to free the world from the tyranny of the atom." He was silent, then, "Yes, we are free to be mystical. We are free to be metaphysical, and it all leads us deeper into immeasurable quandaries. Our ignorance is computable only by the accuracies of science. It is endless."

Nearby, unseen, the ducks were rustling and whispering in their moist mushy voices. They sounded sympathetic and soothing I thought. And Fred continued, "We are caught in a tangling web of non-communication. Nations cannot communicate, they can only organize to destroy each other. By horrible indiscretions of knowledge mankind has released uncontrollable energy that has no conscience. It is possible for it to destroy all that exists—the cities, their art, the humanities the hopes, the decencies of the ages. Toward the black holes, the black holes of space. Are we headed there? Is that the absolute, abominable destination?" "You frighten me." I protested. "Perhaps you are ill and don't know it." "I am very well,' he argued flatly, "but I am stuck in a desert of depression, a desert deeper than the sands of fatal Alamos." I was astonished. It has always seemed impossible for him to speak frankly of his own personal troubles. In the anxiety of the moment I could think of no wisdom to offer, no text from scripture, no inspiration from love or despair. We sat and waited for the intruder, but none arrived.

In the distance an occasional automobile came home and the door was slammed on a day well-or-ill-spent. Twigs and leaves fell here and there from the trees we sat beneath, as nature continued the inconsequential and tidy work of drying and replacing small parts. The wheel of night turned on and on, the stars on the rim moving forward as the dreams of sleepers in the town went with them. Now and then a drift of air brought fragrance from a garden, and all the while the horrible infinity of the black holes hovered above us, omniscience without compassion. I dug my elbow into Fred's side. "We must have some other relationship with the universe." I implored. To my amazement he answered, "What is it?" This was utterly unlike the balanced good spirits and reasonableness of the man I knew. I was convinced that he was ill and some disturbance of body was causing this havoc of mind. "You have read all the philosophers that matter," I said, "and some that are going to be forgotten. Don't they help?" He was silent for a long while, then, "Here we sit," he said, "waiting for an insignificant criminal who doesn't show up. Shall we go home?"

As we stood to leave there was a commotion in the creek nearby, a frantic quacking and splashing. "The ducks!" I cried, and in spite of the soberness of the moment I added, "They fell into a black hole." Fred managed to laugh. "Just a pack of idiots," he suggested.

WE WORKED CONTINUALLY at things of ordinary importance, matters of settling ourselves on the earth we had chosen at cost of tiring effort, pangs of uncertainty, and loss of established home of many years. We had to make some good use of our new land—plant fresh prune trees where old ones stood dying, or let our son tear out the old ones and put in a vineyard for himself. Although our decision to leave our familiar environment and strike out in a new one did not include the rigors and hardships of traditional pioneering, it did include a moral responsibility, a seriousness of association with a new surrounding, even a different attitude as citizens toward the earth. If we had wanted an easier life we ought to have gone where labor was cheap and plentiful. If we had wanted a more exotic life we ought to have gone to some old island of temples and Asiatic art. Or just to Mexico which we already knew and cared for. But it did not occur to us to abandon our own country. The truth is, however, that being too earnest may carry a burden. We could not relax and make too much of the pleasure of our work, we could not be easy-going and behave like jaunty pioneers; only like faithful, weary pioneers. In any case, it took more than light-heartedness to lay down a water system of several thousand feet with all its pipes, valves and fittings, and to put up a deer fence of steel posts and strong wire, and to build a house of complex design and many sheets of blueprints. What Fred had undertaken in retirement was a major work, and being a major person he went at it in a total way. And there were days when the most promising valley promised nothing but more hard work.

Did we need a vacation? Would you like to go to Machu Picchu?

YES!

We went to Machu Picchu and saw its cruel glories, a place whose people must have slaughtered themselves—or their slaves—with hard work.

THERE ARE THE WEATHERS, the season's deputies. At first we thought them not so different from our southern weathers, but later on we began to feel that being five hundred miles closer to glacial memory made us more susceptible to the habit of being cold. Once chilly we could not shake it off. Once wet in torrential rain the weather goes on raining with a vigorousness that pounds down on our hilltop in a noise of deluge and the chanting and trilling of endless frogs. Frogs are maniacs about being frogs. They even come to the kitchen door to tell us that they are frogs and proud of it. As the rain falls on them they twang and tra-la, one hundred at least, full of echoing pride. They hitch up their little wet green pants and shout. We leave the kitchen light turned on because it pleases them. They make love and multiply. We never see a single frog.

Included in the design of our house are, altogether, five areas for plants. Three are "plant rooms"—we would not be so grand as to call them conservatories—provided for plants that are tender and take afront at chill weather in chill latitudes, and two long "plant alleys," incorporated at either end of the house. One has a long table for work, potting, planting seeds, generally making messes, fondling bulbs and entertaining great expectations. The other is for ferns and shade-lovers, and also provides from the bedroom an exquisite view of a pure white flat sasanqua. All of these special areas are carefully intended parts of the house, for my sake. All of them fill me with amazement and gratitude. I know that I am a woman pampered far beyond my deserts, but still aware that what I give to life, however imperfect, is the source of important response in another, creative mind. And for all this I wish to speak gratitude, but find no adequate words, only the feeling of need, much as a plant gropes with its roots for substance in the earth and raises its head for light as day crosses over. I am continually disturbed by the nervous truth of incomplete fulfillments—given what is given to me I should be an astonishing plantswoman perfectionist. I am not. I am only a woman who cries "Stop! Look!" when I am walking or driving and a new flower or tree comes into sudden view, a woman whose memory of the earth is very long, for it was beginning to be mine before I even existed.

It is more than a century and a quarter ago that my paternal grandparents were wading through the swamps and searching the woodlands of Ohio and Missouri, hunting for rare orchids and exchanging recently discovered plants of many kinds with Gray, Wood, Darlington and other masters of early American botany. Ignoring their duties to land and family they would, each afternoon, climb into their saddles and devoutly wander in unexplored country, to shout in reverence, "Stop! Look!" and slide off their horses into the mud. Together with scientists of repute, Henry Beeson Flanner and Orpha Annette, his wife, helped to systematize American botany of the midwest on the eve of the Civil War and the following years of national trouble. The joy of their search and the halo of their excitement have descended on me. To my shame, I lack their knowledge, their specialization and their taxonomy, tired farmers who sat late at night studying Greek, Latin, French, to improve their passion for plants, and the endless assembling of herbariums. I can claim only a little of their ardent brains, as much as can be passed on through filament or stamen and the clot of pollen sticking where it must, tiny and full of life, with life left over.

During many years, and perhaps in fond curiosity, my husband has respected the sap that runs in my veins. He sees, with affection, that I am ignorant. He asks no authority of my mind, but generously acknowledges my need to be a kind of plant.

OVER THE TOP of the hill and part way to the creek there stands the first imposing circle of redwood trees. There are eleven in the outer group, all fine and very tall and many feet in diameter. There are stumps just inside this circle and remains of older stumps still further within, which show that the renewal of life has gone on for a long time and with the ever-restoring conscience of this tree to cling to life and prove its power to defeat death. One day I climbed the road into the woods and to the redwood circle. I was carrying a small light box, tightly wrapped. For many years I had had this box and kept it hidden away; and left it sealed. It was not a box I ever wanted to open, yet it was a part of my life, a box I had to keep. I was alone. I had not asked for a companion. I did not want to inflict upon anyone else the memories that I brought with me; not even upon the nearest person in the world. Now, on a gentle day in summer, at the hour when, far back in the canyon, the thrush was singing like an angel, I followed the road to the redwoods and stood beside them and looked up, far up into their distant spires. I tried to pray, and I tried to evoke from the great trees an answer to the prayer I could not utter. Presently I managed to tear open the tightly closed box, and stepping down from the road to the embrace of the circle of trees, and as nearly as possible without looking, I emptied the box of its contents, tossing them into the enclosure. Then I stood, leaning against the nearest tree. "His body was born dead," I whispered, "But his soul was born alive. It is still alive." I pressed close against the tree, against the solid symbol that endures in the name of life continual. "You know that, my tree. You know that."

And there I stayed and got rid of my tears, and when that was finished I climbed onto the forest road and stood with the shadows and ferns about me, and picked a few of the tiny cones from the beautiful flat branches of the tree.

Far away the thrush went on with his evening melody, those effortless and perfect intervals that filled the woods with soft clarity.

AS WE SAT waiting and watching on the deck on a pleasant afternoon we were celebrating a birthday, a most ordinary yet special measurement of time. Fred was being honored.

There they come! Our family, led by John and Sue with the four grandchildren, walking up from their house to ours through the orchard. Scott, Daniel, Kelly, Lisa. The eldest was ten, the youngest four. In any procession, no matter how quiet and informal, there exists a stateliness, even a haughtiness, that is worth a long wait. This simple procession had formed unintentionally, yet it held a sweet grandeur that moved us warmly as we sat above the orchard and looked happily down while these six people drew near. Human beings moved forward in file, one after the other, in the oldest paintings, on the oldest pottery. They accept their proper sequence down the ages, they follow the priestly or the royal leader. They have a destination. The knowledge that they have a destination carries them forward like the wind in the sails of a fair ship. "Ahoy!" We cried, and the six people all looked up at once, with six smiling faces. It is such a moment of grace that mercifully wipes out the mistakes and humiliations of life which would otherwise devour us.

Another measurement of time stays with me, with both of us, and for its singularity, indeed its eerie consternation, I have long kept it in mind. One night there was a question in the house of our water supply. The spring, our source, was far away, back in the canyon, governed by an electric pump under a redwood tree at the edge of Nash Creek. In the black of night Fred elected to visit the pump and determine why it was not properly feeding the storage tank on the hill. I went with him, and in the faithful four-wheel drive we descended into the darkest pitch of night. If we had been hunting for a darker, blacker spot we could not have found it. With his flashlight Fred worked patiently on the pump. "I'll see if it's flowing better into the tank on the hill," he said. "You stay here." And in a few moments he and the Willys with its beamy lights just piercing the darkness disappeared. The finality of this upward departure was chilling. I stood where I had been left, with my back braced against what I knew was a dogwod tree. Looking aloft all I could detect was one star scarcely visible among the branches of a towering fir. It was not possible to detect anything else. I stretched out my hand. I presumed it was there, yet it was uncannily absent. I felt consumed by the intense element of the dark. I became alarmed about Fred. He had gone into the black night alone. "Take me with you." I cried. Something had happened to him. I suffered a moment of dread. I must set down the memory of it. I must inscribe it here clearly for those who come after me, my dread and my love. The memory belongs to them, it is in the mesh and fabric of their lives. Oh! Suddenly a faint light above, then stronger, loomed on the invisible road, and the four-wheel drive and Fred blessedly, slowly descended, and I recovered my confidence. There was more work on the pump, more departures, and each time the surging return of absolute night.

Then, as I stood with my back against the dogwood tree I was astounded to see, perhaps two yards off to one side, a tiny and brilliant moving light on the ground. Phosphorescence? No, it was positively moving. In fact, heavens, it was moving very slowly in my direction. In its miniscule display it was beautiful, but the alarming pleasure of its company was lost on me. To be alone in impenetrable dark was enough. And then the terror of repeatedly losing Fred up the hill. Now to see the approach of this small, ambulant lantern was not easy to take calmly. On and on it came. Its deliberateness was stupefying, and no less, its seeming purposefulness, as it headed, inch by mysterious cold bright inch, straight for my right foot. I confess that I was foolishly frightened, and that my right foot was terrified. On and on. If it were my habit to recognize such an apparition as a visiting "pixy" or some sort of "fairy"—but no, I do not entertain such explanations. This was a living, ambling gem, some kind of glowing organism, and it had a little glowing will, and it desired to attack me. I was an intruder. Should I scream for help? Fred was up on the hill at the top of the ladder that leaned against the redwood tank. This was my own crisis. If, when he returned he could not find me as he knew me, perhaps he would find a pale, gleaming woman figure, scarcely human, clinging to a tree.

Now the little light had almost arrived, lacking an inch of my craven foot. I closed my eyes. After a long, very long, wait I opened my eyes. The creeping jewel of light was extinguished. Its delicate fuse was out. I understood, and I was ashamed. I was a gross, unillumined creature who had stumbled into the realm of the immaculate, the formidable flake of burning frost. With flawless dignity it had withdrawn, just before it soiled itself by touching me. Without a flash of fear it entered the pure black pit of time.

I HAVE READ THAT place has memory. People have memories; places, too, have memories. If one finds it difficult to agree with this at least it accounts for the phenomenon of haunting which has existed, accepted or denied, for a long time. We live in an age in which the real and the unreal face each other in the public eye: or is it chiefly in the eye of poetry, which has become as many-faceted as a fly's, far more stern and endlessly reflecting. In any case it is a time of splendid confusions and we are constantly challenged to believe more than we can comprehend. But I find it easy to believe the strange matters that I am about to relate.

We had only recently come here to live. We are not the first people to live on this land. The Indians had been here for four thousand years, and the white man a hundred years on these acres before we came. We were just settling into our new home. It was summer, and a black night. The moon, soon to rise, was not yet shining. I was sitting in the house without any lights turned on, feeling the animal dark all around me and wishing to feel it even more, as a thing I could touch. The night was very still, when suddenly I began to hear a noise. It started on the right side of the house out of doors and slowly moved across the garden, close by, toward the left. It was a heavy noise, the thudding of hoof-beats, a horse's feet slowly plodding, and at the same time the loud creaking as of a farm wagon being pulled along. It went as far as the woods and ceased. I listened for a brief while further, and hearing no more, I sprang up in the dark and hurried to turn on a light.

At that moment Fred appeared and asked breathlessly, "Did you hear it?" I answered, "Yes, I heard a horse pulling a wagon." He looked at me incredulously. "Now what are you talking about? I mean, did you hear the scream?" "Heavens, no. Tell me." He said, "I was standing by the fence along the woods in the dark. I don't know why. Something drew me there. And just then I saw the merest glimmer of light and knew the moon was about to show through the trees, and thought, 'I'll wait a moment and see it.' We always enjoy the sight. I took a single step toward the fence and right there, full in my face and right at me was this awful, I tell you, terrific, scream." "Was it a scream of pain?" He answered, "It was pain, anger, hatred."

There is no more to our story. Nothing has ever occurred since to be added to it. We are the once-haunted, and not again. The moon rose as usual above the dark woods, and if it looked down on agony long ago, the place thereof has kept its secret.

ON TOP OF THE HILL in the woods Fred has discovered a spot of rich earth where he has planted a fine garden of vegetables. It is strange, it seems to me, that such good earth is perched aloft on the hilltop. But there it is, and edenic at the edge of the woods, and in his care it produces very fine beans, beets, tomatoes, carrots and all the assembly of most virtuous vegetables in great quantity. Day after day, spring, summer, autumn, he rides noisily and slowly up the steep road in the four-wheel drive and works at his garden task. Sometimes I join him to help, perhaps to pick the emerald beans that hang gracefully from the bamboo tepees he has erected, or to direct the stream of water from the redwood tank just above. I love this hill's crown where oak and madrone spread their boughs above far views of the valley.

Fred has built a deer fence and they never find their way in or over. The jack rabbits are charmed to stay out. For hours at a time in this sylvan place the gardener works and contemplates. He must be happy here. I like to think that he is very happy. I wish poignantly that I had joined him more often. Coming down with boxes and buckets of vegetables just severed from the plants in exuberant freshness he makes a grand sauce of green peppers and tomatoes and basil that will preserve the summer months on the hill in gorgeous color and aroma for the dark winter.

The road to his garden leads on down to the creek and into a grove of redwood trees. Enormous trunks, they rear up into a sky that seems to hoist itself in order to give them the space they need for their height. Tall as they are, however, they are second growth. It was here, in woods now owned by us, that Sam Brannan, developer and exploiter, had cut timber in the late 1850s to build spa, hotel and bar and establish in mud and affluence the town of Calistoga. He is revered as pioneer and founder. We have the stumps he left behind and the circles of restored life which even intoxicated entrepreneurial muscle could not wholly demolish. Back from the creek a few hundred feet is one huge stump, more huge than the rest, with its top high above the ground level. When that majesty came down the earth must have been convulsed while rocks bowed low to trembling ferns. Sam Brannan was a tall rich man. Given his footage and wealth, the one six feet and three inches, and the other a full million, it is still hard to understand how he could have the nerve to lean back on his heels and point up and yell to a lumberjack, "*Take it down!*" Today the stump's flat top is capped by a playhouse which John has built for his children. A ladder takes them up where they can view their surrounding realm of tranquil wilderness. They are too young to despise Sam Brannan for his destruction of native grandeur. They enjoy the mossy elevation provided for them by a nice guy.

MADIA, blooming at the wrong end of the year and the wrong time of day, madia, whose aroma clings like a soft bur of fragrance to the sleeve of your coat as you brush past, madia is slight and subtle in its meaning, and that way it lasts. Its memory is one that holds expectation. As day finishes I watch for the gold corollas to open around the disks of maroon, and I lean across the plants and breathe a new evening, a new air of evening. It is difficult to say why this weak flower and its phantom scent are so real, so significant. Yet madia is, for me, one of the haunting revelations that meaning, memory and roots are small things, light enough to hang in a spider's lace, not heavy enough to tear the silk of that elastic fringe.

Perhaps easier to accept for memory and its reasons is the bright lemon yellow of the intense maripose tulip much earlier in the year. It is an infallible yellow, a concentration of radiance and purity. Yet in the brevity of its blooming and the mere sprinkling of numbers, it, too, is a frail reason for memory. But reasons have nothing to do with the fact. The facts are the small things that locate the heart. One of the best was when the little brown creeper flew through the fence from the woods and clung to my knee, thinking me to be a tree. As he looked up with trust into my human face my delight at his confidence, my love for this mite of a bird, were so intense that I was for that spontaneous moment profoundly at home, I was joyfully rooted.

There are other things, totally different from nature. There is my sister Marie's fine piano. I remember how I had it shipped with much apprehension from New York to Southern California when she died, because my son said that some day he would like to have it himself. I remember how he despised his piano lessons. I remember, with a kind of torment, how I had it brought north with us, and somehow dragged up the narrow twisting road to our house. I myself cannot play it, this excellent Steinway, except with the most trying imperfections, yet this very frustration is itself a kind of root. As such I accept it. Oh, yes.

Memories are roots. Roots are small, fragile, webby, big, crazy unreasonable, ephemeral, wistful. They are what we are made of. Even our bodies, for there must be a place for our souls to be contained. Never mind what my name is. Let's use your name. We are all here together. We are at home.

And there are the roots of loss and sorrow. Having these grim roots, we are no longer strangers in the land.

HUMAN INTELLIGENCE through the centuries has evolved miraculous skills of mind and matter. It has cleared the ancient labyrinth of its stealthy destinations. It does not hesitate to deny an absolute axiom, and thereby discovers delirious truth and a new century. It is a mysterious failure of the human intelligence that it has given poor service to human beings about to face unknown events ahead of them, even tragedy, without knowing it is imminent. Human intelligence, by which we live and believe in the next day, has never evolved a nerve, a strategy, a system of private turbulence, a shock of blood and fear to warn us when tragedy is about to explode at our feet.

Mere anxiety—the impulse to worry—is a habit so common, unreliable and wasteful, that wise people try to ignore it. It rarely tells the truth, the human truth. Truth is what we are unutterably in need of, stark, painful truth, hard as biting teeth, sharp as many needles. Then we would know in time what we would have to know and what human intelligence, perfecting its objective skills down the centuries, fails to tell us before the devastating moment when we find it out for ourselves.

WHAT IS TOLD HERE can be spoken only in incoherence, in tears and sobs of disbelief. Yet grief demands dignity, and tragedy makes its announcement in bitter calm, for it must be heard. After that, sorrow, which is all there is now, can take over.

So it was, that on a morning in the month of October I came from one end of our home, where I had been occupied with insignificant chores, and found my husband lying dead, still supported against the counter in the hall, as if he had tried to catch himself in falling. It was half an hour since we were last together and if he had felt ill he had not said so. I shall never know whether, in the desperation of his last breath, he called out to me for help, and I was beyond hearing and gave him no help, but now threw myself down beside him and cried out to him myself for help, and implored in anguish, "Take me with you!"

AND NOW I ARRIVE at the dark, empty present, alone, without you, without the light and warmth of your closeness, without your smile, your smile so full of your eyes. To what unchartered borders of years and earth have your life and mine hurried and passed beyond, always cherishing each other? By count of the calendar it has been a long way. By count of love it has been brief beyond believing. And now by count of loneliness, brief beyond enduring. The years have gone by, searingly quick, and searingly short. Your death is too bewildering to accept as truth. I cannot believe in your non-existence. Then you must still be here, but in this engulfing, drowning silence, where? Why, everywhere, tantalizing my desperation, my solitary state. Memories by the score, magically disturbing, spring from the mind that grief at once plows deep. The hardest to bear are those of remorse, the things ill-done, or never done at all. You must be hearing what I must say. Don't go away. I cherish unspeakably all that rushes into my mind as my memories return. They are good, they are bitter, I must respect them all, joy and grief, pride and humiliation, enchantment or shame.

We know that there are only two people in the pages of this book. If they seem to disappear, they return. Their dialogue is chiefly with the earth, but the earth only seems to be listening, is that not true? The earth of wind, rock and water takes no notice of men and women, while men and women faithfully make poetry, meat and wine, and living art from the earth, and from where they lie in each other's arms, cry out in passion to their indifferent host and repeat the words that Eve spoke to Adam, which must have been, "I am sorry. Forgive me."

It is common in our day to speak of love with irony or guarded commitment. Only in sad honesty can I say that an extraordinary man is dead, and that I am an old woman, alone. "Get out your diaries," you said, "Let's read your diaries." "Yes. Yes." Let me open them today at hyacinth and gold of a Mojave sunset, or vivid companies of spring flowers as they storm the hills back of Santa Ynez, or the red hills and blue lilies of Paskenta. Let me open the pages I wrote long ago and read to you now aloud and without restraint or shyness what the embrace of joy was like to two young lovers. And what more, what more? Always your talents of creation, your accurate hands touching paper, and copper plate, the impressing of your vision and design upon the space waiting to be filled. And more? Always more. The gathering of a few excellent works, that we might live in the presence of other people's skill. What more? The concern for this valley always shared. And it is endless, and always more.

The indulgence of grief is poor behavior. The day will come when I am hungry again, and ordinary life will impose itself upon me decently. I will sit at our old cherry-wood table and raise my glass to your empty chair. I will say, "Congratulations, my darling, for a good, quick death," and nobody will guess that I am still in anguish on the floor beside you.

More, ever the earth we have known together, from the first entries in old diaries through all our trips and the many years in a semi-tropical garden of the south, and once on a glorious cruise that took us all the way by Pacific waters up ten thousand feet to Cuzco and Machu Picchu. Together we stared at marvels of Peruvian gold. Together we hated the Spanish conquistadores. There is no end to remembering.

There are, as well, the sounds of this land, the sound we love best, the soft and vibrant voice of the quail. Dear bird and friend, are you not the very legend I searched for as I stood and looked across the valley? The familiar image that warns and reassures and leads to safety? You with the forward tilting and attentive plume, pause and say *Yes,* say that I am not alone where you run in your smooth stride of hastening, as if on little nimble wheels, across the earth of my western garden.

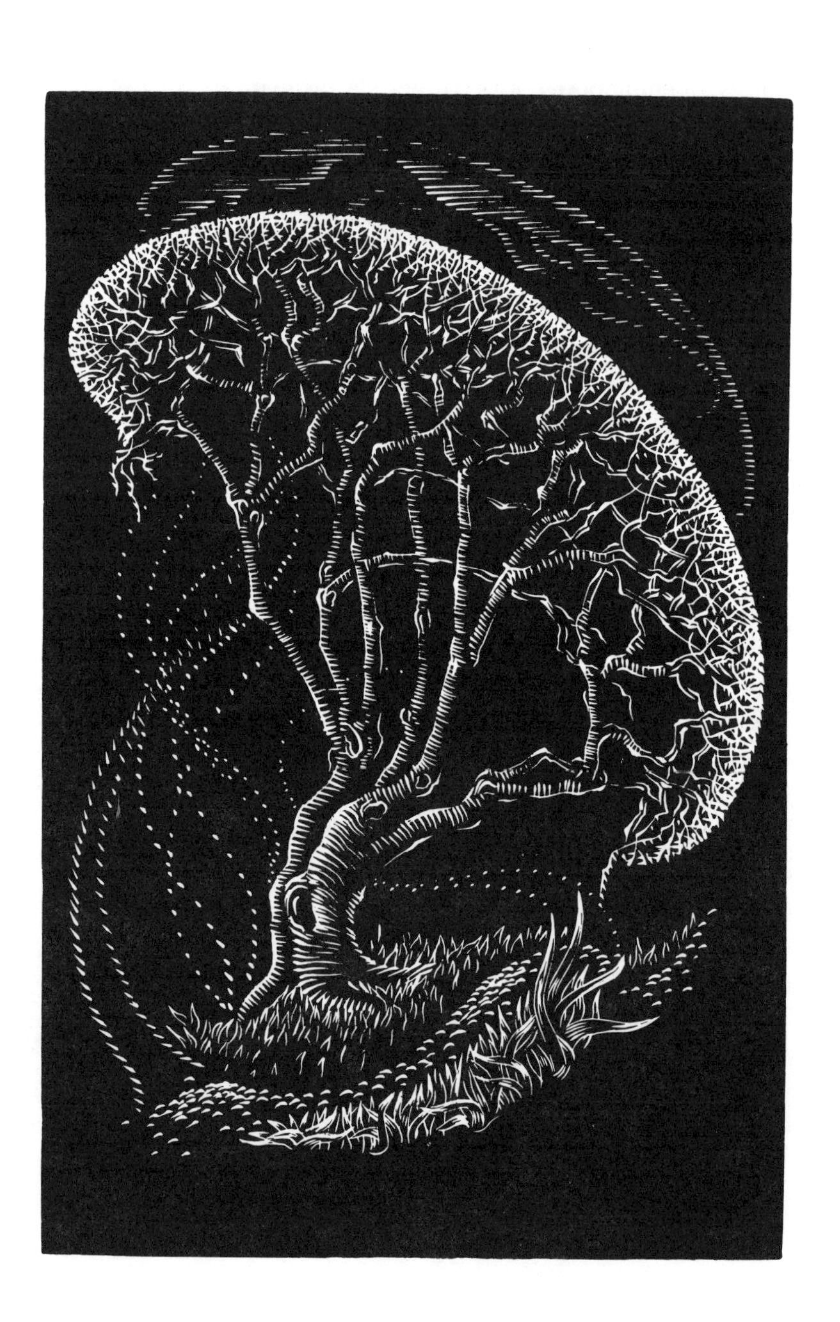

THE OLD CHERRY TREE

TOGETHER WITH a few human beings, dead and living, and their achievements, trees are what I most love and revere. In my life and my concerns I have defended trees that are threatened and praised them when they are ignored; for their sakes I have made enemies of friends and neighbors. The knowledge that in a western canyon I am the owner of very tall straight redwoods and firs gives me pride of citizenship beyond anything I might accomplish by my own efforts. I have taken a serious kind of joy and a delightful kind of peace in their shade, and at night I have watched the shaggy white planets pass above their dark branches. I have loved trees, I have planted trees and have been excited to grow a tree from seed and discover the first minute sign of unfolding life that will, some distant year, become a rooted tower or a spreading bower of rustling foliage. I am fortunate to live where native trees are numerous and where horticulture is popular and every rural family has an orchard.

Our own orchard is old and contains many trees that are dilapidated, but even the most dilapidated have been safe for years because each time I look at them, aware of their crookedness and awkward appearance, I also see some odd curve of bough or improvisation of flowering that make sudden poetry of them and of their trashiness. They are thus protected, although they should, under good management, be sensibly discarded and replaced. Yet the more dubious they look the more they resemble the paintings of Oriental masters and suggest the fresh enduring emotions of the ancient anthologies. Perhaps there is universal truth in implying that real meaning is wrung out at the last moment, and the last moment must be prolonged. Whose duty is that? Really not mine. Yet I have discerned it. I am involved in its total meaning, of tree and of human. Still, as I myself grow older and, alas, much older, the image of new young trees in place of the old ones creeps temptingly into my mind. It is a troubling image and its disturbance, just now, settles on the spirit of a woman wandering and trying to think in an old orchard.

If I could have the sense and courage to take down a decrepit tree in spite of its fanatical habit of reminding me of the brave words of General Su Wu written down 2,000 years ago as he sadly embraced his wife for the last time, and in spite of its mad annual impulse to bloom and bear and spill bushels of amethyst prunes and rosy apples on the ground, it might be for me a spiritual rejuvenation and consequently even a beneficial thing for the flesh as well. But conscience is tyrannical. It is a womanly vice. I have become the guardian not of my own but of whatever other life remains in the earth I possess. It is hard to be wise and natural. Although my husband is no longer living I can guess what he might wish to do. To remove a tree that had lasted too long and replace it might well please him. This thought is a poignant incentive, also a consolation for the choice I foresee as melancholy and difficult, whenever it must be made.

The decision I am afraid of, on a mild February morning of our western spring, is one for advice from a clinical priest. It is beyond horticulture. It becomes moral. In my mind it faces and fights itself and I am torn. Must I continue to identify myself with the aged, no longer serviceable and eccentric trees, or do I dare to relate myself to young trees with futures, good looks and green chances of tonic sap? My orchard, by usual standards, is conspicuously shabby. This hurts me. And as I look around I notice a few unrewarding cherry trees.

To get a cherry in our orchard has meant to rise early, before birds or worms are awake, and snatch at a fruit or two as day breaks. One summer, in order to enjoy the large luscious black Bings, my husband ordered a nylon net he saw advertised in an agricultural supply magazine. The advertisement promised protection to fruit from marauding birds and complete accessibility for the picker. If all of this was true, we were not the ones to demonstrate it. As we hoisted the net on poles to the top of the tree the clinging mesh snagged on every twig on the way up; then, as we squinted into the sun and gave helpful and confusing suggestions to each other on how to free the net to slide down as directed, we found that it could descend only by being profanely and scrupulously removed from twig to clutching twig. However, the picture as given of happy people congratulating each other outside the net while the cherries waxed abnormally big and the birds fretted on the next tree—in the advertisement they were scowling—this fine fiction was worth keeping in mind until suddenly and utterly we tied ourselves in. The next step was just to ripen along with the cherries in boredom and frustration, while the birds jeered nastily. This very personal recollection comes back to me as I find myself standing under a cherry tree, not the one that lassoed us, but a very old pie-cherry tree, largest and surely the oldest inhabitant of the orchard. Indeed, where is there in the entire valley a cherry tree so old and so big? It is a tree known in the neighborhood, and when we first came here to live friendly strangers drove up our hill and requested, "Just a few cherries, please, from the old tree, enough for the wife to make one pie."

Now so much of the tree is fallen or dead that there is scarcely enough fruit for a robin to make one cherry tart. I get none myself, even on tiptoe or a ladder. And the trifle of birdsong it holds is nothing to give regard to. Neither has it grown ancient with picturesque aspects. All of its gifts are gone. Even the child with a little basket passes by. It is too large, you see, to be forgotten, yet neglect is its fate today. And so I talk to myself as I look up and see no buds swelling toward wide-open blossoms where the bees should soon be rolling. From something so unpromising, what is to be expected? Why is it so difficult for me to say, "It is time, old cherry tree"?

This happens to be the day on which my son is preparing a level space where there will be built a storehouse for tools and equipment. I welcome this plan as one to maintain order. Not only his powerful tractor, but various archaic automobiles considered by him too beautiful and valuable to go to the dump, can now be respectably housed, particularly that very sacred hot-chili-red truck that usurps the sight of grace and elegance where my tallest bamboo has established its feather culms. This is a day of orchard and premises keeping and its purposes begin to take hold of me. I look up at the cherry and assess what I see. Crippled and lumpy, here and there split, beheaded of several heads, and to all appearances so nearly dead that there seems no way to say it is alive. Then I look up at the sky and straighten, as if to think of other things. A good clear morning to be alive myself. And John has opened the deer gate and is driving the loud clanging tractor into the orchard toward the work that makes ready for his storehouse.

"Oh, John," I call. He doesn't hear above the roar of the machine. Without thinking I reach out and touch the subject I was about to speak of. I touch not only a tree of bark and wood but with tingling certainty in my fingers I touch an entire century. At this spot there stood a house which no longer exists. The spring that served its occupants still serves us from its cold stone trough at the edge of the nearby woods. And a county clerk had written the first deed to this property, dated 1878, in script almost too lace-like to read or believe. It was then that the tree was planted and began to work for men and birds and until this spring it has never stopped working. It was always the first to bloom and ripen, always prompt no matter that the weather might delay, still it brought on its sweet sparkling globes of fruit. For 100 years of faithfulness there should be a reward. Hang the old tree with garlands, strike up the fiddle. But I am caught up in another momentum. Again I call, "Oh, John!" He can't hear, there is too much noise. I go closer. "Maybe it's time," I shout, and he yells, "What is it?" How hard it is now for me to be positive and loud, quick and wise. But I must not take time to be careful. I am committed. "Time to take down the old cherry tree," I shriek.

I know that decisions like this should be made in quietness and deliberately. We should live slowly, even timidly, in imagination with all the possible results of the irrevocable. Once down, there would never be an up to this deed. I exercise a frightening power. It is not exactly a choice between life and death because life is already attacked by mean and obvious details of the end. However, the power of termination is awesome. Naturally, it chokes me. I cough. Kings, tyrants, judges—how have they arrived at that last fatal word that condemns life without feeling their own lives threatened, shredded and about to come apart? But don't be silly, I say to myself. What are you talking about? Just get on with what there is to do. Isn't this the mistake you have always been making? Too much emphasis on the wrong thing while you let the right thing drift? "John!" I shriek again. "Time to take down the old cherry tree!"

"Yes, yes," he says easily, "I can hear you," and I become aware that he has turned off the noise of the tractor and also that in the interval of silence I can hear the demented sound of the ranging peacock that forages in the foothills and ravines near our place. It always seems to be a sound of mental stress and at this moment it is right for my state of mind.

I am surprised by hesitation on John's part. "I don't think I can cut it down, it's too big," he says. "I'll have to push it down." Again he seems doubtful. "If I can."

And the tractor starts again. Then for a while he is busy with his work of leveling nearby, but as I stand watching he begins to look at the old cherry tree in a calculating way. I suspect that he welcomes my decision to get rid of a tree so dominant yet unproductive in this place, and he must be astonished. He circles and comes closer. I cringe and brace myself for the shock to my nerves and my conscience. John backs his tractor and then goes forward with an awful clang. He hits the old cherry with a loud dull crash. It does not budge. It does not even quiver. He backs away. Again he charges forward and collides with the tree. It stands without shaking. It is only I who grow weak. I shake and feel sick. Perhaps I am wrong to condemn a creature so full of strength in spite of the many signs of being done with strength. Again the tractor charges and without effect. I would prefer to leave now, to go back to the house and hide behind closed doors where I could neither see nor hear; but it was I who started this turmoil and I must remain to see it through.

At this moment my 13-year-old grandson, Danny, arrives and stands near me, watching. "Your father is wasting a lot of good gas," I scream. "Diesel," he screams in return. "Fuel," I scream back at him, as the attack continues. Backing and charging, backing and colliding, the tractor roars and hits, and at last the old cherry tree begins to tremble. But it stands. I suffer and watch in a nauseating agony of indecision. Should I save the body while some secret obstinacy still holds it up? Or is it better to let the savage shocks continue? Suddenly John begins a new maneuver. He backs, then closes into the attack at an angle, tilts the sharp, wide blade of the tractor and digs it into the earth. Again he does this, and again, until the action of the bellowing machine and the chaos and the frightful uproar become a kind of violent choreography. I stand riveted and overcome by what I have started in the quiet orchard. Finally the tilted blade snags on a massive root. The root holds.

The contest goes on. My son will not give up. The old tree will not give up. The tractor will not give up. The boy and I stand and wait and the diabolic ballet goes on and on. The tractor, although a monolith, has been converted into a maniac of circling and twisting power. In its fierceness it is serpentine. At last, at last! With a heavy snap, a sound of fatal resignation, the root breaks. And the tree still stands! The tractor attacks it for the final time. And when the old cherry tree falls there goes down with it a century of hopes and many kinds of weather, sun and drought, of good rain and poor rain, of good pies and poor pies, and 100 years of countless round white flowers opening and coming apart and drifting down while early-rising and late-loitering birds and bees came and went and always, at the right cloistered moment there was the invisible sap slowly storming up through the trunk and into the tips of the branches, just as it was rising in a million other trees at the eternal hour given for ascension.

While I stand and have no voice to speak John nonchalantly gets down from his tractor and walks over to the fallen tree. He begins to peel off patches of thick bark. "Termites," he says. I do not want to see them. "More termites," he announces. I hate the sight of them. I stay where I am. Then he pulls off another, larger patch of bark. "Look here!" he cries. There is a company of small lizards, eight in all, spending the winter in the shelter of the cherry tree. John collects them and quickly puts them all down Danny's back. The boy rolls his eyes and draws up his shoulders but does not flinch.

I retrieve the young saurians and put them in the grass. Then I go to the prostrate tree to hunt for more, as if this intimacy with small creatures might reduce the magnitude of the old giant's final resignation. There seem to be no more lizards, but I find neat necklaces of empty holes, the precise work of woodpeckers whose echoing labors I have often heard. Then John, in the cavity left nearby, finds a beautiful little snake of a dark rich skin, a young gopher snake whose presence close to the tree appears to indicate dependence and community. John holds the snake for a moment and we watch as it flashes its rapid tongue in the sunlight. Then he puts it down and we see it take off, small and solitary. We are left alone with the fallen tree. It is stripped of everything, except its misshapen size, and its weighty bulk so lately upright and adamant. There it lies.

John gets his power-saw and starts methodically cutting branches for firewood. So soon does the drama and ordeal of destruction become the routine of plain use. Dazedly I pick up a few small logs. "Are you out of wood at your house?" he inquires.

"No," I answer, "This is ritual. I want a few pieces of the old cherry for my bedroom fireplace."

He tells me, "It won't burn yet. It has too much sap."

"Too much sap!" I exclaim and hastily drop the wood. "Is it still alive?"

"Well, what do you think? You saw how the old girl fought back."

His personification of the old tree horrifies me and I begin to cry.

John gives me a well-controlled look. "You're queer," he comments.

With difficulty I inquire, "What's the diameter?"

"Three feet at least, I guess. Big for an orchard tree." And he goes to work again.

In wretchedness I pick up the pieces of wood and hug them, small logs of smooth bark ornamented with delicate silver-green medallions of lichen. I carry them ashamedly to my bedroom porch. When I lay them down I know that I will never burn them, ever, no matter how long I keep them to lose their sap. They are too elegant, they have too much meaning. I shall never wish to warm myself at their melancholy and accusing blaze.

I return to the orchard. It is a still, empty place now. John has driven his tractor back through the deer gate, and Danny has gone with him. I stand and stare at the remains of the old cherry, the limbs in a heap, thick bark strewn, the powerful roots split and twisted. Now no one else will know if I give in to tears as I realize that there must always be new questions in my mind about the imperfection of my decision to remove the old tree. I have learned, too late, that there is more to life than what is visible. The greater strength had been underground and out of sight, and I had grossly, stupidly, not even guessed it was there. You fool, you made a wrong choice and you only proved that decisions are hell, a fact you've known since tormented childhood when it was not possible to be sure whether vanilla or strawberry or chocolate was the right choice, or whether to wear the sash with rosebuds woven into the silk or the blue satin one or the one with Roman stripes. I observe how tough the roots are and how strong and sharp and that they point up with a kind of spiraling hiss into the placid noon sky. How dreadfully eloquent they look, expressing all that I felt for them. It is not easy to stand alone with them.

Something prods me into an attempt to understand that the moment holds a finality beyond agitation. In weariness I can only decide that an urban-minded person would take this with helpful sanity, and I shall never be an urban-minded person. "For you they pulled out oak, fir, madrone and manzanita," I say to the roots. "It was a long time ago. Do you remember?"

For many minutes I stand here where the first axe wound and the first gouge of the plowshare cut into this very ground at my feet where the old cherry tree has just been knocked over. Now I hear the tractor again. It is down in the vineyard. "All right," I say, "It's true that I am queer. I talk to trees. I talk to roots. It's true they can't answer. But they have a lot to say. Look at them!" And I myself look at the roots where they lie on top of their trash, full of fierce power in every slashed point that thrusts up, full of a cherry tree voice, full of a forest voice, the forest that fell to make room for an orchard 100 years ago.

I start back to the house. "Just to get out of earshot," I tell myself.

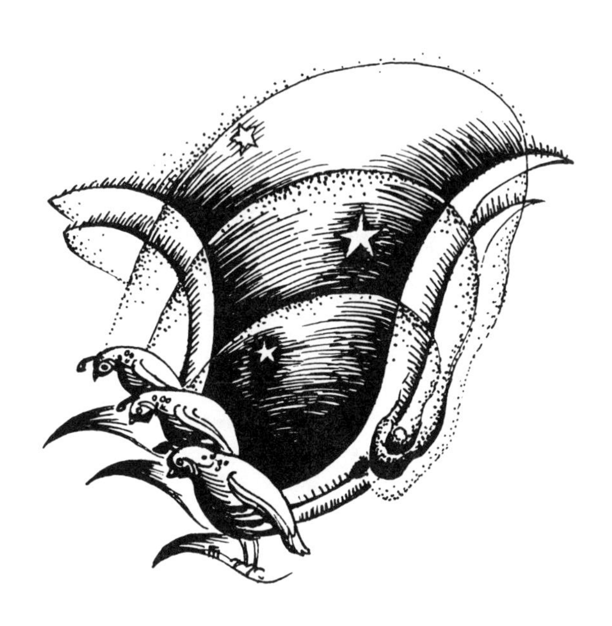

ABOUT THE AUTHOR

HILDEGARDE FLANNER was born in 1899 and has lived most of her life in California. Her books have been published by Macmillan, Porter Garnett, Grabhorn Press, New Directions, Ahsahta Press, Ikuta Press (Kobe, Japan), and No Dead Lines.

Her articles and poems have appeared in *The New Yorker, Westways, New West, Botteghe Oscure, Poetry, The New Republic, The Nation,* and a host of literary quarterlies. Her forte is essays and poetry of western provenience, with particular devotion to plants. According to Diana Ketcham of the *Oakland Tribune,* "Hers is the generation that knew how to write. The unpretentious grace of Flanner's prose should make us regret she hasn't published more."

The essays in BRIEF CHERISHING concern the years Hildegarde Flanner spent in the Napa Valley of Northern California with her husband, artist and architect Frederick Monhoff.